ENVIRONMENTAL BIOLOGY AND ECOLOGY

Laboratory Manual

Third Edition Revised Printing

Les M. Lynn Ph.D.

Bergen Community College

KENDALL/HUNT PUBLISHING COMPANY
4050 Westmark Drive P.O. Box 1840 Dubuque, Iowa 52004-1840

Revised Printing

ISBN 0-7872-6145-9

Printed in the United States of America
10 9 8 7 6 5 4 3 2

PREFACE

For over twenty years, my chief frustration in teaching an environmental biology course for non-science majors has been the lack of viable laboratory exercises that can be accomplished in a three-hour time period. When consulting other lab manuals, the exercises have been field-oriented which prove difficult in an urban-suburban setting. Some labs have been more ecologically oriented with a tremendous emphasis on statistical analyses; sometimes above the heads of non-science majors. Some labs necessitate access to the laboratory at all times of the day or night-difficult to do in a commuter school.

After years of modifying labs, which have included a tremendous amount of busy-work and filling in the three hour period with audio-visual material during which most students became comatose, I have finally attempted to write a manual in hopes that lab time will have some significant meaning. Only my students and the others who use this lab manual will be able to judge whether I have been successful or not.

I have included 14 lab exercises plus a field trip in this manual which follow the syllabus to my course, Introduction to Environmental Biology. While some of the labs are geared to the specific conditions on the campus of Bergen Community College, and pertain, in some cases, to northern New Jersey, it is hoped that they can be sufficiently modified to be meaningful in any college setting. Since they are true laboratory exercises, it is possible for anyone to use them.

For the instructor, Appendix A contains the materials needed for each lab. I have also included annotations to each lab in Appendix A, which hopefully, will help the instructor. The range of lab exercises necessitates a working knowledge of many fields of biology. As one reads these exercises it will be evident that some labs have been modified from other disciplines (e.g. microbiology) so they are relevant to environmental biology. Other labs span the realm of classical ecology, botany and zoology. Many of the exercises use materials that we have constructed on campus. It is hoped that by the time this manual becomes available, the special materials that are necessary for some of these exercises are also available. If not feel free to contact me through the publisher.

I hope that these exercises keep the student occupied for the entire laboratory period and spark a curiosity into the environment around them.

There are many people who I would like to thank. Many thanks go to Lynn DeMarigny and Sharon Sawey, faculty in the Division of Natural Sciences and Mathematics here at BCC. Our labs could not run at all without the great job of these technicians. They are directly responsible for the construction of some of the customized lab materials that have been conceived on this campus. The great art work is the product of Ron Fox. His contributions and ideas in making this lab manual a reality can't be minimized. Thanks goes to Glen Van Olden of the Passaic County Soil Conservation District for materials included in this manual. I would also like to thank my wife Blanca, my daughter Monica and Butchie, the wonder dog for their understanding during the writing of this manual.

Finally, to those students who are still asleep after participating in the previous lab experiences; it was their snoring that finally gave me the impetus to write this manual.

Les M. Lynn
Paramus, New Jersey
January 1993

PREFACE TO THE THIRD EDITION

This third edition has really developed an identity. There are now 20 laboratory exercises and one field trip. There should be more than enough material for a semester's work in a non-science majors environmental biology course.

What I have tried to do in this edition is clean up some of the exercises, that is, remove what doesn't work, modify existing exercises and try to remove the busy work. I have found that requiring students to draw endless pictures of slide material has proven to be tedious for both the student and myself. Unless the student has some artistic ability, their drawings are usually worthless. There is some drawing but I have reduced the quantity.

The identity that this manual has assumed naturally reflects my personal bias in that it is heavy on ecological exercises but put in the context of letting basic ecological principles explain and illustrate the conflicts between humans and their environment.

I have deliberately refused to include exercises such as at home energy audits and the like. Going home and counting how many light bulbs one has in one's house, is not very appealing or while visiting a sewage treatment facility might be worth while, the logistics of such a field trip may be impossible. I have tried to get students to understand ecology, environmental biology and to do some real science: gather data, analyze the data and draw conclusions from their results.

The labs in the second edition have utilized the computer and in this third edition; two of the new labs not only utilize the computer but also data downloaded from the Internet. I believe that this is, in part, the present and future of science for non-science majors and that our concepts of "laboratory experience," must change to reflect new technology and the changing backgrounds of our students.

The preface to the first edition thanked all of my students who fell into deep sleeps attempting to accomplish the endless busy work that the first twenty years of this course required. After four years of using this manual, while it is not perfect and will continue to change, I believe it is pretty good. To all of my students who are now alert but are faced with full three-hour labs, I apologize but this is the way it should have always been.

I would like to thank my wife Blanca, my daughter Monica and Butchie the wonder dog for their help in preparing this manual.

Les M. Lynn, Ph.D.
Paramus, New Jersey

August 1996

TABLE OF CONTENTS

Page

Name_________________________ Microscope #_______

EXERCISE 1: THE MICROSCOPE - A NON-TRADITIONAL APPROACH

Many of the lab exercises that we will do this semester entail the use of the microscope for the entire, or at least part of the lab. For the non-science major, learning the "theory" and the principles behind how a microscope works takes the fun out of its use. For this lab it will be more important for the student to be able to use the microscope properly rather than be able to explain how it works. By the time the student leaves the lab he or she should be an expert microscopist

Using and caring for the microscope properly will be a key objective of this lab. These scopes are expensive and precision instruments. They should be treated as such. The methods for teaching this lab are fool-proof. If the student follows the instructor's directions implicitly, by the end of the lab period, the student will be able to use the microscope in all the applications for this semester's work.

If the student doesn't understand the instructions, ask the instructor, not another student. If you do not know what to do, ask!! If the student has already mastered the use of the microscope in another course or school, BE PATIENT!!!! The instructor will be teaching this lab at a very deliberate pace. Don't jump ahead. The instructor will only ask you to stop and this will slow down this exercise even further. Simply put, don't do any more than what the instructor asks for and don't do any less. If you do not know what to do, ASK THE INSTRUCTOR!!!

STEP 1: THE SET UP

When your instructor calls your name, get a microscope from the cabinet. Please, tall people take scopes from the top shelves so that the more diminutive individuals can take them from the bottom shelves. Carry the microscope with two hands, one around the base of the scope and the other on the arm. Don't swing the scope as you go back to your seat! The scope has a number on the back of the base, either tell your instructor the number of your scope or fill in the number on the proper form on the instructor's lab table. Then, COPY THE NUMBER OF YOUR SCOPE IN YOUR LAB MANUAL, IN THE SPACE PROVIDED ON THIS PAGE!!! Place the scope on your desk and do nothing. Wait for further instructions.

Before using the microscope, we must familiarize ourselves with the major parts of the scope. After this is done, label the diagram of the microscope on page 8.

STEP 2: THE PARTS

The major parts are:

a) Ocular Lens(es) - the eyepiece(s).
b) Objective Lenses on a Rotating Turret, low power, medium power, high dry power (some scopes have a fourth objective, the red-banded one called the oil immersion lens, which we do not use in this course). Twirl the turret. The objectives "click" into place. You can feel them "click." Now, "click" the shortest objective, low power, into place.
c) Stage Control Knobs are usually found on the right side of the scope and toward the front. Move the top knob and the stage moves up and down. Move the bottom knob and the stage moves left and right. In this manner, the microscope slide can be positioned or centered perfectly on the stage.
d) Microscope Stage is where the slide sits. Note how the stage clips hold the slide in place.
e) Coarse Focuisng Knob is the "big knob." It is located both on the right and left sides of the scope and is used to initially get the slide in focus. If the knob is turned, the entire lens mechanism of the microscope moves. For now, turn the knob so that the lens are as far down as they can go.
f) Fine Focusing Knob is usually in front of the coarse focusing knob (or in some cases, nested within it). When it is turned, you cannot see the len mechanism move but it is. This adjustment fines tunes the object and brings it in to focus clearly.
g) The fine and coarse focusing knobs are located on both sides of the scope, but on the left side of the scope is a third knob, the Condenser Adjustment Knob. Turn it and note that the condenser, which is directly under the stage, and is seen through the center hole in the state, moves up and down. Move the condenser to its "up" position. The condenser "straightens" the bean of light as it comes up through the scope. Light has a tendency to scatter and the condenser, minimizes this, directing maximum light up through the scope. Why do we need a condenser? Some of the objects that will be observed will be stained (dyed) and will require maximum light. Other objects may be translucent and require that the amount of light be minimized so that they can be observed. A great deal of living material falls into this category. Look at the condenser itself. The condenser has a lever (some have two). Move the lever to the left and right and notice what happens to the amount of light coming through the condenser (look through the "hole" in the microsocpe stage). As the lever is moved, the light coming through the lens increases and decreases. This is the Iris Diaphragm.
h) Iris Diaphragm works like the lens of a camera, regulating how much light comes through the microscope. This is a second method of adjusting the light.
i) Light - turn the light switch on and off and leave it off for now.
j) Base - bottom portion of scope.
k) Arm - supports the lens mechanism.

UP TO THIS POINT YOU SHOULD **NOT** HAVE BEEN LOOKING THROUGH YOUR SCOPE.

Before using your microscope, even before turning the light on, all the lens and glass surfaces should be cleaned using the lens paper that the instructor demonstrates. Clean the following parts:

a) ocular lens (eyepiece-either monocular or binocular)
b) objective lenses on the rotating turret - usually three or four lenses. The smallest objective lens (low power) may be recessed and cannot be cleaned.
c) the condenser lens - move it to its up position.
d) the light source - clean this before you put the light on because it gets very hot, very quickly.
e) It is a good idea to clean every microscope slide that you use.

STEP 3: USING THE SCOPE

Obtain a slide of the letter "e." The "e" can be seen in the center of the slide, it is not the "e" on the label! Without looking through your scope place the slide on the stage with the "e" over the hole in the stage where the light comes through the condenser.

Now, slowly, using the coarse focusing knob (big knob) focus upward until the "e" comes into focus. If you do not see the "e," notify your instructor (check to see if the low power objective is in fact, "clicked" into place). Now using the stage position knobs, center the "e" perfectly. Make sure it is dead center in your field of vision. Now, with the fine focus knob, fine tune the focusing of the "e."

In review, you have focused and centered the object under the microscope. If you have the "e" focused and centered you can now move the medium power objective into place. This is usually accomplished by rotating the turret COUNTER-CLOCKWISE until the next largest objective lens clicks into place. Now look through the ocular and using the coarse focusing knob get the "e" focused.

The "e" should be focused with a minimum of adjustment-it should essentially be in your field of vision. This is due to the fact that the microscopes we use are PARFOCAL. This characteristic of the scopes simply means that if you are centered and focused under one power, the object should be centered and need a minimum of focusing under the next higher power.

WHAT SHOULD YOU DO IF YOU CAN'T FIND YOUR OBJECT UNDER MEDIUM POWER? (HINT: ASKING THE INSTRUCTOR IS ONE SOLUTION BUT WHAT IS THE OTHER)?

How does the "e" look under medium power? How does it differ from low power? Under low power, how is the "e" different from just looking at the slide with the naked eye?

After focusing under medium power with the coarse focusing knob, center the object exactly and using the fine focusing knob, fine tune the object.

The "e" has now been focused and centered under medium power.

We are now ready to switch to high power and as long as you are focused and centered under medium, you can now swing the high power objective into place. Make sure it "clicks" in.
DO NOTHING AT THIS POINT. DO NOT TOUCH A THING!!!!

Note how close the high power objective is to the microscope stage!! If you were to start focusing with the coarse focusing knob, you would send the objective lens into the slide, smashing it!!!!! This would result in serious **repercussions that are horrible and agonizing.**

SO UNDER HIGH POWER, YOU NEVER, EVER USE THE COARSE FOCUSING KNOB. ONLY USE THE FINE FOCUSING KNOB. ALWAYS!!!

Some of you may not have the "e" and those of you who do have it will only have a small portion of the "e" under the scope because we have magnified it so large (remember that is the point of using a microscope - making small objects larger!).

What should you do if you can't find it under high power? The key to finding the "e" under high power is making sure it is centered perfectly under medium power. If you still can't find the "e" under high power, ask your instructor for help.

THIS IS A FOOL PROOF METHOD FOR USING THE MICROSCOPE. IF YOU ARE FOCUSED AND CENTERED UNDER LOW POWER, YOU CAN MOVE TO MEDIUM POWER AND IF YOU ARE FOCUSED AND CENTERED UNDER MEDIUM POWER, YOU CAN MOVE TO HIGH POWER WHERE YOU NEVER FOCUS WITH THE COARSE ADJUSTMENT KNOB!!

STEP 4: PRACTICE, PRACTICE, PRACTICE.

How do you get to Carnegie Hall? The answer is just above. The way you become comfortable with using the microscope is the same way-practice.

The following three slides are of protozoans, microscopic creatures that live in slow moving streams and ponds. These creatures are stained (dyed) but when you get to high power you may have to reduce the light (use the condenser adjustment knob) to see them clearly. Regardless of whether these creatures focus in sharply or not, start on low power, focus and center, switch to medium power, focus and center and then switch to high power, focus (fine focus only). You may do these creatures in any order. They will be studied in detail later but for now they are only to give you practice in using the microscope. The creatures are: Amoeba, Vorticella and Paramecium.

THE PROTOCOL IN THIS LAB WHEN DOING MICROSCOPE WORK IS TO ONLY TAKE ONE SLIDE AT A TIME. DO NOT TAKE STACKS OF SLIDES WHICH WILL SURELY FALL AND BREAK. NO EXCEPTIONS. If a table of four wishes to take four different slides and exchange them among themselves, that is alright.

STEP 5: JUST A LITTLE THEORY

First of all the function of a microscope is to make tiny objects larger (rocket science, huh?).

The eye piece (ocular) is usually at a magnification of ten power (10X). The three objective lenses are 5X, 10X and 43X (low, medium and high, respectively). So we calculate the total magnification by multiplying the ocular power by the objective power. You do it.

Low:

Medium:

High:

What do the lenses of the microscope do to the object? Think of what the "e" looks like to the naked eye and what it looks like under low power-rhetorical question-the microscope reverses and inverts the image. The lenses make "right," "left," and "left," "right," and up is down and down is up.

STEP 6: ANOTHER EXERCISE.

Many times we will be making our own slides of living material. The procedure is called making a wet mount. Place a **SINGLE** drop of water on a clean, blank microscope slide. Place the object to be observed in the drop (unless the object is already in the drop). Carefully, and at an angle, lower a cover slip over the drop and gently place it on the slide. Voila'! You have made your own slide. The important concept here is that living material usually is translucent- without color. You must reduce the light coming through the microscope using the condenser adjustment knob or iris diaphragm.

Make a number of slides of "hay infusion." You should observe a variety of "creepies" and "crawlies" swimming around.

STEP 7: ONE MORE EXERCISE!

We will make a wet mount slide of one of our own cells, called simple squamous epithelium. These cells make up tissue that usually serves as a lining of an organ. Simple squamous epithelial cells line your cheek (of the mouth, not the other cheek). Prepare your wet mount as per the instructions in Step 6, but use a drop of the stain, methylene blue, instead of water. Gently, with a clean toothpick (one per student, please) scrape the inside of your cheek and place the residue in the drop of stain. You only have to scrape gently, you don't have to draw blood!! You will have these cells on the tip of the toothpick even though you may not think so. Place a cover slip over the drop and work the microscope from low to medium to high power. In the space below, draw a diagram of these cells under high power. Note that the cells are irregularly geometric but have a round nucleus. Label the nucleus.

Repeat the same procedure using a drop of water and the tip of an Elodea leaf. Elodea is a submerged aquatic plant. Locate the rectangular cells and find the nucleus for each cell. Draw a number of these cells below.

What are the obvious differences between the plant cells (Elodea) and the animal cells (simple squamous epithelium)?

STEP 8: PUTTING THE SCOPE AWAY

All kidding aside, these scopes must be put away properly, no exceptions. It's simple and straight forward but in your haste to finish up you may forget.

1. Take the slide off the stage and return it to the proper tray.
2. Swing low power in place.
3. Rack the scope down so that low power is close to the stage.
4. Wind the light cord around the base.
5. Return the scope to the cabinet and place it where you found it.

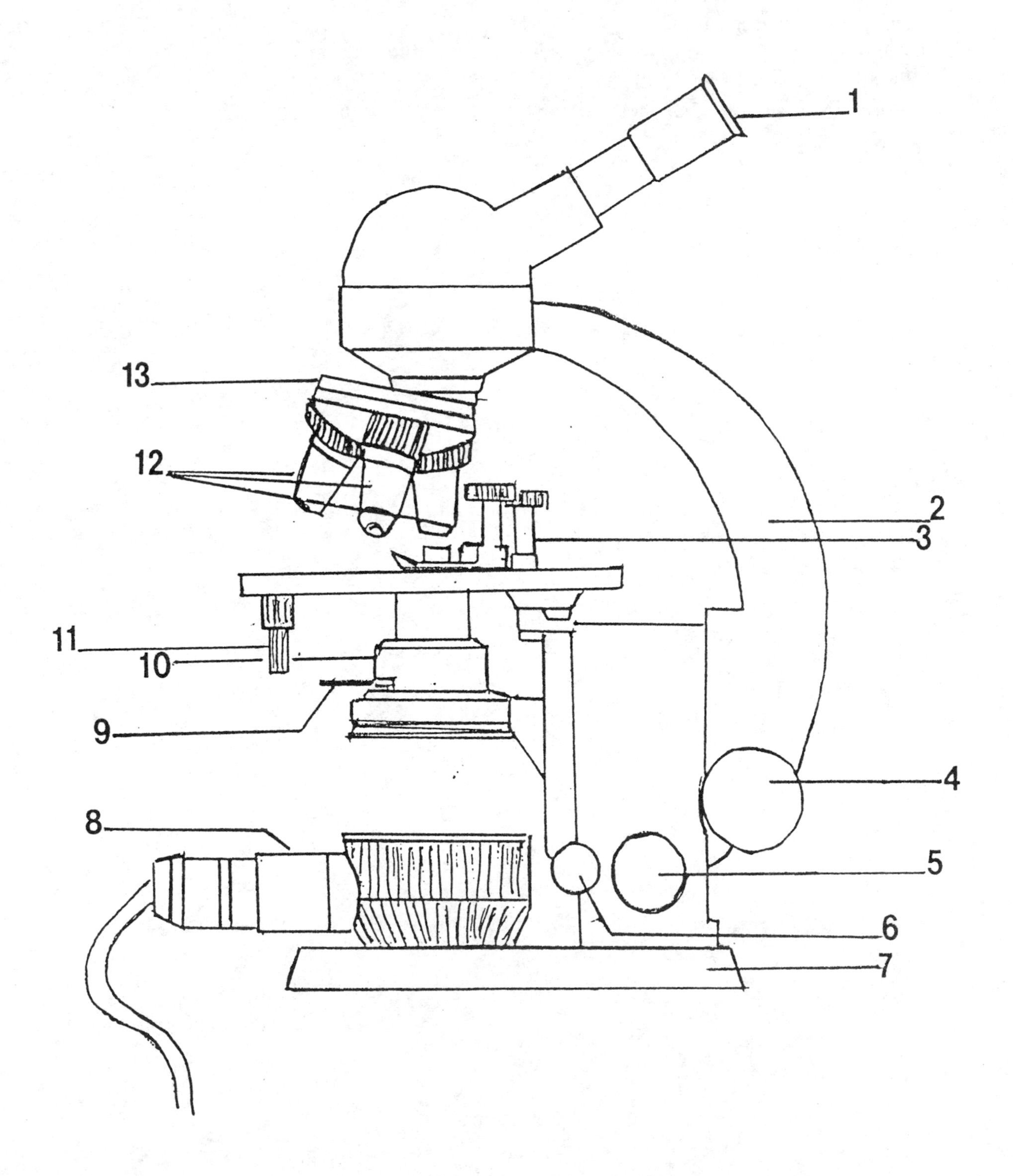
1
2
3
4
5
6
7
8
9
10
11
12
13

Name___________________________

EXERCISE 2: AN INTRODUCTION TO ECOLOGY

Ecology is the science that studies the interactions between organisms and their living (biotic) and non-living (abiotic) environments. It is the study of nature's structure and function. In the first part of the lab, we attempt to reinforce some of the more important concepts concerning the science, in a variety of activities.

The basic unit of study in ecology is the **ecosystem:** an area where all of the organisms interact with their physical environment to produce a flow of energy. The flow of energy in an ecosystem leads to the recycling of materials comprising that ecosystem, species diversity (the ability of the ecosystem to support a varied community of organisms), and defined trophic levels (specific ways in which organisms obtain their nutrition). The flow of energy in an ecosystem supports organisms that obtain energy in very particular ways.

The components of an ecosystem can be divided into the abiotic components, the non-living components, and the biotic components, the living components of an ecosystem.

The abiotic components are factors such as the sun, climate, weather, substrate - that which an organism lives in or on, and the minerals and nutrients that are necessary for living organisms, collectively referred to, in this connotation, as "inorganic materials."

The biotic components are the living organisms and can be divided into the "producers" and "consumers." The producers are organisms that can produce their own food, within their bodies. These are the green plants, which undergo photosynthesis. A more accurate term for these organisms is "autotrophs." All of the other organisms are consumers or "heterotrophs." They must consume other organisms (living or dead). Plant eaters or herbivores are primary consumers because they eat the producers directly. Secondary consumers eat primary consumers, tertiary consumers eat secondary or primary consumers and in rare instances, an ecosystem can have a fourth consumer, a quaternary consumer that eats the tertiary, secondary or primary consumers (all are carnivores). Other consumers are the decomposers, bacteria and fungi that break down dead creatures into their constituent elements and are crucial for the recycling of substances within an ecosystem. Detritivores are creatures (such as earthworms and some insects) that eat partially decomposed material - **detritus**.

For the first activity observe the aquarium and list the abiotic constituents of this ecosystem and then list the biotic components of the aquarium. Attempt to classify the organisms by their trophic levels (e.g. producer, primary consumer, secondary consumer, etc., etc.) Look out a window of the lab. What type of ecosystem is out there? Do you get the idea that an ecosystem is an area, a specific place, but it does not have set boundaries? It can be as small or as large as whatever area is being studied. Repeat the activity as for the aquarium: list the abiotic and biotic components of the ecosystem that is outside of the laboratory. Although you may not observe all of the components for either ecosystem, **include those characteristics that might be present - use your imagination!!**

AQUARIUM:	Abiotic Characteristics	Biotic Characteristics

GOLF COURSE:	Abiotic Characteristics	Biotic Characteristics

There are two "ecosystems" in terraria that should also be categorized. The woodland terrarium contains a representative collection of organisms that might be found in the forests in this area. To get an idea of what the woods are like, read the material concerning the temperate deciduous forest in the next section on biomes. Then as done previously, list the biotic and abiotic components. Construct a simple food web using the information you have learned during this exercise.

Bogs are ecosystems that will be dealt with in some detail when wetlands are discussed. Briefly, they are wetlands characterized by an accumulation of peat moss, low oxygen and in many cases low pH (acidic conditions). The waterlogging, low oxygen, high acidity, etc. do not promote a great deal of microbial action-decomposition. The peat accumulates. In the case of ombrotrophic bogs, water only enters the ecosystem from precipitation and leaves by evaporation (there is no inlet or outlet to the bog). Bogs form on the margins of ponds and lakes where these characteristics are present and given enough time, will fill the lake in completely. The vegetation must also deal with a rigorous environment: low nutrients, high acidity, low oxygen, water-logging, etc. As a result, some of the plants have special adaptations for surviving. Some have become insectivorous. The well-known Venus flytrap comes immediately to mind as well as the pitcher plant and sundew. Sphagnum moss grows on the surface of the bog with many herbs and shrubs growing out of the sphagnum. Heaths such as blueberries, huckleberries, leatherleaf and Labrador tea dominate the shrub layer. Under the living layer of sphagnum is the accumulation of dead sphagnum-peat!!

Animals are usually transient in bogs; they pass through but rarely take up permanent residence.

As in the previous exercises, draw sketches of the representative bog components and construct a simple food web based upon the information in this exercise

BIOMES

A biome is a large geographical area with a specific type of climate, and as a result, a specific biota (community of plants and animals). Historically, it was the classification of large geographical units based upon both the flora and fauna as opposed to one or the other or by climate alone. Regardless, there aren't many biomes on the entire planet!! The number varies (10-20) depending on the classification system used. In the United States the major biomes are: temperate deciduous forest, boreal coniferous forest, desert, tundra, grassland, woodland and a few others, again, depending on the classification system used.

The **temperate deciduous forest** is found in eastern North American (as well as Europe, Asia and South America). The climate is characterized by the four seasons. There are many mitigating factors within the temperate deciduous forest that determine the biota (fire, soil moisture, other microclimatic differences, etc.).

The temperate deciduous forest is subdivided based upon the dominant vegetational types. In New Jersey, the forest is called the former oak-chestnut forest. The dominant trees are oaks and formerly, American chestnut. A fungus, the chestnut blight, accidentally introduced at the beginning of the 20th century from Asia, has effectively wiped out the chestnuts (they are not completely gone but only grow to saplings before the blight kills them). They are no longer important components of the forest. Chestnut has been replaced by more oaks, maple, hickory, birch and hemlock.

North of the former oak-chestnut forest is the mixed mesophytic forest of the Catskills and central New York. This forest type also dips into the most northern part of New Jersey. Dominants are American beech, sugar maple, tulip tree, hemlock and others.

The fauna of the temperate deciduous forest includes white-tailed deer, opossum, Eastern chipmunk, red fox, Eastern coyote, black bear, three-lined skunk, box turtle, timber rattlesnake, copperhead, leopard frog, bull frog, spotted salamander, many birds and many other organisms. There are at least seven other subdivisions of the temperate deciduous forest.

Along the Atlantic Coastal Plain and in localized areas in the Northeast, are pine-dominated forests. On the eastern tip of Long Island, New York, in the Albany, New York area and occupying much of southern New Jersey are the "pinelands" or pine "barrens."

In New Jersey, these forests are dominated by pitch pine, which is fire resistant. While the understory vegetation is destroyed by the many fires in the region, pitch pines are resistant to all but the most serious "crown" fires. It is this ecological factor, **fire** that perpetuates the pinelands. Without fire, oaks and other species would replace the pines. In the New Jersey Pinelands many rare species of both plants and animals are present and found in few other places.

Two other biomes need mention and they are biomes found at high latitudes and elevations: Arctic tundra and the boreal forest (in Europe it is called the **taiga**). **Tundra** is the biome found above the tree line in North America, Greenland, Iceland, Europe (Scandinavia and Russia) and Asia (Russia). This biome is *circumpolar*! The biota is very similar throughout the tundra. Growing seasons are very short, 100 days or less. Winters are long and very cold. All of the plants must accomplish their biological activities (growth, reproduction, food accumulation, etc.) within this period. There is little precipitation in the tundra. Heavy snowfall would be assumed but this is not the case. Precipitation is between 4-10 inches per year!! The snow just blows from one spot to another. Although the soil thaws during the summer, it is just at the surface. **Permafrost** characterizes the tundra. As a result, drainage of water is very poor. Soils are waterlogged and the tundra is distinguished by thousand of square miles of boggy habitat, the **muskeg**.

Herbaceous plants dominate tundra vegetation: grasses, sedges, mosses, lichens and herbs. There are few shrubs, which are dwarfed, and no trees. Animals include rodents like the lemming, as well as caribou, wolves, arctic fox, arctic hare, snowy owls, polar bears (very descriptive names!!), musk oxen and many others.

South of the tundra, also circumpolar, are the vast coniferous forests of the north-the **boreal forest.** The boreal forest is characterized by long, cold winters, short summers and while overall precipitation is low (16-40 inches per year), there are heavy snowfalls.

Dominant trees are the spruce and fir conifers. Cedar and tamarack are also components of the forest. While a shrub understory is sparse, there are many herbaceous species on the forest floor. Animals are represented by moose, deer, grizzly bears; cougars and wolves are the major predators. Many rodents, including mice, squirrels, and porcupines are present. The Canada lynx and its chief prey organism, the snowshoe hare, are two characteristic animals of the boreal forest. Wolverines, fishers and other weasels are present.

Elements of both the tundra and boreal forest appear at more southern latitudes but at high elevations. It is possible to observe elements of the boreal forest in the Adirondacks and Catskills of New York State and even in the Great Smoky Mountains in Tennessee and North Carolina. Alpine tundra, representing the herb and shrub zones can be found at these higher elevations, above the tree line, as well.

Deserts occur in the arid (dry) regions of the planet. They are usually found along cold ocean coasts, in the rain shadows of mountain ranges or in the interior of continents. In North America, they are found in the southwestern United States and on the eastern side of the Rocky Mountains. Worldwide, there are deserts in northern and southern Africa, along the coast of Chile in South America, Australia and Asia.

Rainfall is very low (10 inches or less per year) and evaporation rates are very high. Deserts can either be very hot or in higher latitudes, very cold. The vegetation is characterized by its sparseness. Plants are usually widely spaced from each other, are spiny (as a defense against herbivores) and to some degree, succulent (fleshy-storing water). Animals must be well adapted to living in such rigorous environments. Many are nocturnal, active at night and inactive during the day when it's the hottest. Many have developed physiologies for retaining all their moisture so they don't have to drink and produce a solid nitrogenous waste (as opposed to waste containing water, e.g. urine). Mammals are represented by the peccary (a wild pig), coyote, white-tailed deer and many rodents. There are reptiles and some amphibians.

For this exercise, draw sketches of the components of the desert, boreal forest and tundra biomes. Then using the material just observed, and your knowledge of these biomes, construct food webs for them in the space below. Characterize each component as to its trophic level (e.g. producer, primary consumer, etc.).

In this exercise, one of the most important tools of an ecologist will be utilized, the **Powers Of Observation**. While some ecology is done in a laboratory or sitting in front of a computer, many ecologists still go into the field to study organisms or ecosystem characteristics directly. Many times ecologists go into the field with preconceived notions as to what they are looking for and depend on their powers of observation to accomplish their tasks. In many instances, it is impossible to study every individual or every aspect of a particular ecosystem so the ecologist must institute a sampling regime. In sampling, what is hoped for is that the representation of the entire ecosystem (or other unit of study), **the sample**, is accurate. But there are always errors in sampling. Later in the semester we will do a number of labs dealing with sampling ecosystems (without leaving the laboratory!). For now we will concentrate on improving our powers of observation. As the activity continues, there should be an improvement in your powers of observation.

The last four pages of this exercise have two illustrations on each page. They appear to be identical however, there are at least six differences between the top picture and the bottom one. List the six differences for each pair of pictures in the spaces that are provided. The only hint is that there are no color differences between the pictures (any differences in color are due to the printing process).

If time allows, one or more of the excellent videos, concerning ecology and environmental science will be shown. Take notes if it is appropriate.

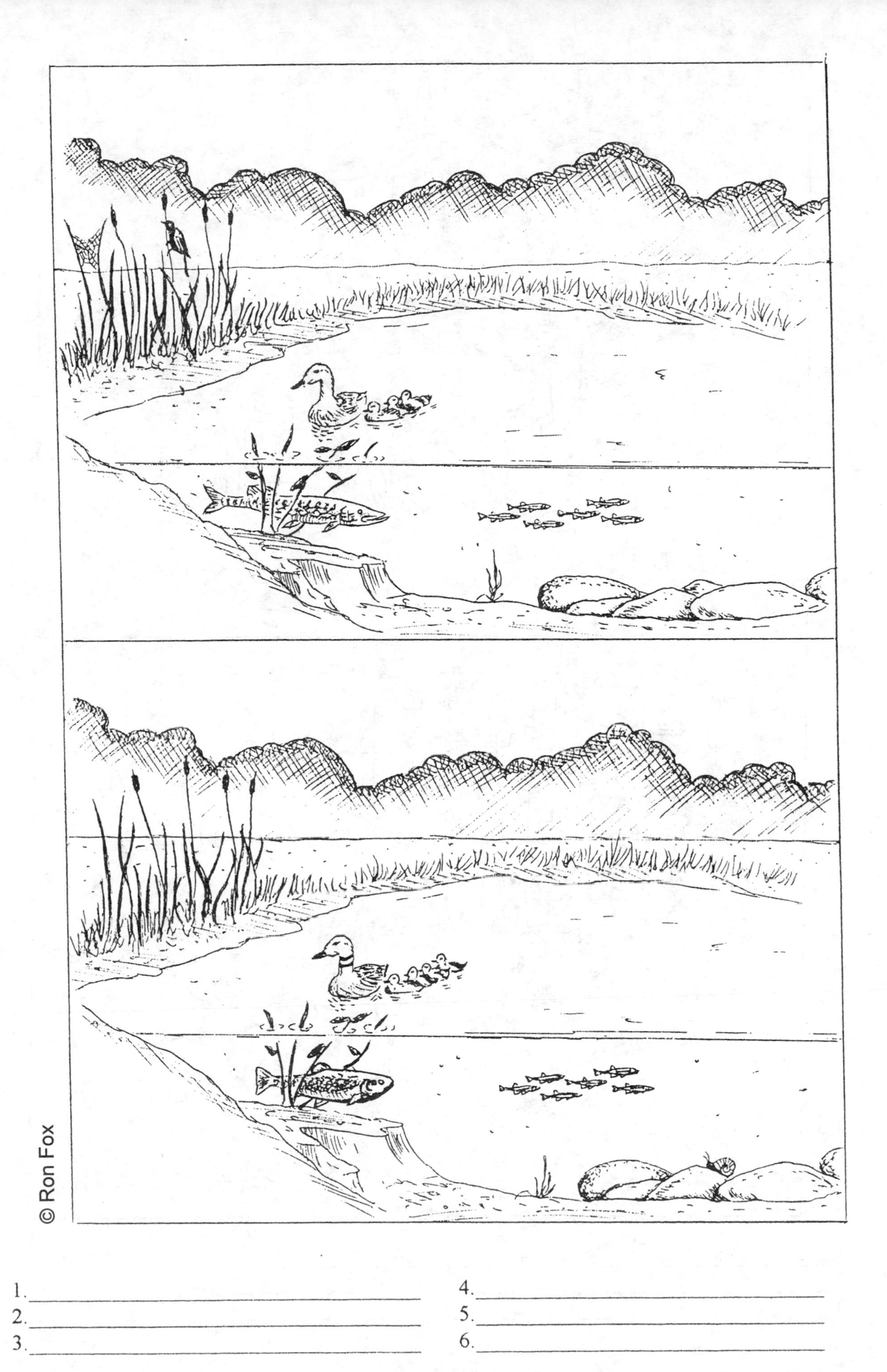

1. ______________________
2. ______________________
3. ______________________
4. ______________________
5. ______________________
6. ______________________

1. ______________________
2. ______________________
3. ______________________
4. ______________________
5. ______________________
6. ______________________

1. ______________________________
2. ______________________________
3. ______________________________
4. ______________________________
5. ______________________________
6. ______________________________

1. ____________________
2. ____________________
3. ____________________
4. ____________________
5. ____________________
6. ____________________

Name____________________________

EXERCISE 3: POND STUDY

A good exercise to improve microscope skills and to learn about an important ecosystem is examining the microscopic life in a pond. Depending upon the time of year, a great diversity of living organisms can be observed during this exercise.

Microscopic organisms that are suspended in water, with little or no ability to swim are called plankton. The plankton is divided into the "plant-like" phytoplankton and the "animal-like" zooplankton. While many organisms live in the water column, rising and falling in response to environmental cues (such as light), it is more likely to encounter organisms when they are attached to debris such as rocks and growing plant material.

In the first part of this exercise, a number of prepared slides will be observed and sketched in the spaces provided. These organisms represent many of the different groups of living organisms and may be found in the pond water that will later be examined. Please remember that the protocol of this lab is to take only one slide at a time, do not take a "stack" of slides. Your sketches are not expected to be works of art however, they are not expected to be scratches on the paper. Take some care in making your sketches but do not spend inordinate amounts of time on them.

Below are the organisms on the prepared slides followed by a very short description of the groups into which they are categorized. The magnifications that they should be observed under are also included. Your drawings should be at the magnification which allows them to be sketched and observed in the easiest yet complete manner possible.

	MAGNIFICATION
I. Phytoplankton	
A. Blue-green algae	Low, Medium, High
1. *Anabaena*	
2. *Oscillatoria*	
B. Green algae	Low, Medium, High
1. *Spirogyra*	
2. *Oedogonium*	
C. Diatoms	Low, Medium, High

II. Zooplankton

A. Protozoa	Low, Medium, High
1. *Amoeba*	
2. *Paramecium*	
3. *Stentor*	
4. *Vorticella*	
B. Hydra	Low
C. Planaria	Low
D. Rotifers	Low, Medium, High
E. Arthropods	
1. Cyclops	Low, Medium, High
2. Crayfish	Low

The blue-green algae are now called the Cyanobacteria because they are more like bacteria, which are very primitive rather then like algae which have a more complex cell type. The blue-green algae (e.g. *Anabaena* and *Oscillatoria*) are usually associated with detrimental algal blooms (population explosions) which are undesirable and degrade fresh water ponds and lakes. Blue-green algal blooms are usually signs of eutrophic conditions, whether natural or cultural.

The green algae are the most common type of algae and occur naturally, in great abundance and diversity in ponds and lakes. They can but not always bloom, especially in the summer months. While not usually indicative of a degraded environment, when there are green algal blooms, usually eutrophic conditions are responsible. *Spirogyra* is an interesting alga because of the shape of its chloroplast (organelle which functions in photosynthesis). Can you guess the shape? *Oedogonium* (the "O" is silent) is interesting because it is an alga that has separate structures for the production of two different gametes (sex cells). Look for them.

Diatoms are desirable algae and come in many beautiful shapes and designs. They can bloom in the spring and fall and are associated with the turn overs of a lake during those times; this is a natural phenomenon and does not infer a deteriorated environment.

Protozoa are the smallest of the "animal-like" plankton. It should be noted here that classifying and categorizing the algae and protozoa has been a controversial topic for a long time. Botanists have wanted to call these organisms plants because many of them are photosynthetic. Zoologists have wanted to call them animals for their "animal-like" characteristics. Ernst Haeckel, a German biologist in the nineteenth century, basically reconciled these problems by putting these creatures in a kingdom of their own. Rather than the animal or plant kingdom, Haeckel created a third kingdom (at least at that time; there are 5 or 6 kingdoms, depending on who you listen to, today) called Kingdom Protista.

Amoeba is the "blob" from the horror movies. It moves by "oozing," actually by what are called pseudopodia. *Paramecium*, *Stentor* and *Vorticella* move by using cilia, small hair-like structures that move in synchrony. While Paramecium is found swimming about in the water column, the other three are usually attached to debris in the water or on rocks, plants, etc.

Hydra is a wonderful little organism that is a relative of jellyfish and corals. It is basically sessile (not moving), attached to rocks and debris but is an active feeder, using tentacles with stinging cells to capture and kill prey.

Planaria is a flat worm living on rocks. It has very rudimentary organs that are light sensitive - "eyes." It also has both sets of reproductive organs so it doesn't have to date.

Rotifers are microscopic organisms swimming in the water column. While they are slightly more advanced than the planaria, they are merely members of the pond food chain, eating debris and in turn, being food for larger animals.

The arthropods are an extremely diverse and numerous group of organisms including the insects, spiders and their friends, lobsters, shrimp, crabs, etc. Cyclops is a relative of the lobsters, etc. but is microscopic. It is an important part of the food chain. Crayfish are the freshwater counterparts of lobsters, which are marine. In the South they are harvested and cooked up to make quite a delicious meal.

For the second part of this lab, prepare wet mounts of the pond water. Try to get some of the debris because that is where the organisms will be found. When an organism is found, try to identify it or call your instructor over to identify it. Remember these organisms are not stained so they will be more or less transparent. REDUCE THE AMOUNT OF LIGHT COMING THROUGH THE MICROSCOPE USING THE CONDENSER AND/OR THE IRIS DIAPHRAGM.

Make numerous wet mounts and record the organisms that are found on the blackboard as well as draw sketches of them in the space provided.

PHYTOPLANKTON

ANABAENA

OSCILLATORIA

SPIROGYRA

OEDOGONIUM

DIATOMS

ZOOPLANKTON

AMOEBA

PARAMECIUM

STENTOR

VORTICELLA

HYDRA

PLANARIA

ROTIFERS

CYCLOPS

CRAYFISH

WET MOUNTS OF POND WATER

PHYTOPLANKTON

ZOOPLANKTON

Name___________________

Exercise 4: PALEOECOLOGY

Paleoecology is the reconstruction of past environments. Using the fossil record, both abiotic and biotic environments of the past can be reconstructed.

The term "fossil" not only refers to animal remains, but plant remains including seeds and leaves (macrofossils) and pollen and spores (microfossils). Algae, zooplankton and even beetle remains can be used as they are contained in the fossil record and give evidence of past environments.

The science of paleoecology depends upon the acceptance of the **Principle of Uniformity**, that is, the processes that occurred in the past are the same processes that control the planet at the present and will in the future, and must be accepted as true. In paleoecological terms, if we know what plants and animals existed in the past, we can **infer** the climatic and other environmental conditions that allowed those organisms to exist at that point in time.

For the past twenty-five years, global warming, the "greenhouse effect," has become of greater and greater concern. It may be the most serious global environmental problem of the present and future. Computer models - simulations, predict what may happen to the planet sometime in the future when CO_2 concentrations double from pre-industrial levels. These predictions of future global warming, are based upon the data of the past - paleoecological data. Using the Principle of Uniformity, we can predict the future by knowing what occurred in the past.

This laboratory exercise will utilize data gathered from numerous paleoecological studies which examined fossil pollen, the male gamete of a flowering plant.

Over the last two million years the western hemisphere has seen major fluctuations in climatic conditions resulting in four glacial periods, each separated by warmer interglacial intervals.

A glacier is not easily defined. Generally, it is an ice mass existing mostly on land and consists of impacted snow which has crystallized into ice, and refrozen meltwater. A glacier develops when snow accumulates year after year (it doesn't melt entirely during the warm months). Due to increasing pressure from new snow and ice atop the accumulation of previous years, the glacier begins to flow. The most recent period of glaciation, the Wisconsin glaciation, resulted in a glacier that covered virtually all of what is now Canada and the northern United States. The ice sheet was thousands of square miles in area and in some places, between one and two miles high!

When the climate cooled (by approximately 5° C), glaciers formed in the northern latitudes and spread southward, covering the land and obliterating everything in their paths. As the climate fluctuated, so did the advance and retreat of the glaciers. This resulted in the change in the distribution and abundance of plants and animals. Animals could migrate away from an advancing glacier. The plants were obliterated but their reproductive disseminules (seeds, spores, fruits, etc.) were blown or carried away. So the plant species did in fact migrate away from the glacier. When the glacier retreated, the propagules, from new plants existing in refugia (refuges) migrated back to their former ranges.

About 120,000 years B.P. (B.P. - "before the present," which is counted back from 1950), the last glacial epoch began with the spread of the glacier in the Laurentian Mountains of Canada southward (this became known as the Laurentide Ice Sheet). By 18,000 years B.P, the glacier had reached its maximum extent, covering all of New England, New York (except for what is now Allegheny State Park, in western New York), and northern New Jersey. The glacier covered New Jersey south to about what is now Asbury Park and west through Denville. By approximately 14,000-15,000 years B.P., the climate had warmed and the glacier had retreated north so that New Jersey was ice free. As warming continued, the glacier continued to retreat. By 10,000 years B.P., the last remnants of the glacier were in what is now Wisconsin, hence we refer to this glacial advance, as the Wisconsin glaciation. This glacier was really the result of two ice sheets converging. The largest covered the entire east, the Laurentide Ice Sheet and the smaller, covering part of western Canada and the Pacific Northwest, the Cordilleran Ice Sheet.

The interval of glaciations, from 2,000,000 to 10,000 years B.P. is associated with the geological period known as the Pleistocene Epoch. We are primarily concerned with the end of the Pleistocene (late Pleistocene) and the period from approximately 10,000 years to the present, known as the Holocene (recent) Epoch.

How is pollen used as a paleoecological tool? As the glacier melted and retreated, depressions filled with water and created lakes, ponds, bogs and other wetlands. The plants that grew around these systems deposited some of their pollen into these basins. Pollen is naturally resistant to decomposition because of the nature of the cell wall. Anaerobic conditions in these aquatic ecosystems further retarded the decomposition process. So pollen is deposited in chronological order, from the oldest (at the bottom of the lake bottom) to the most recent depositions (at the top of the lake bottom). By using a coring device, the sediments containing the pollen can be retrieved in the reverse order in which they were deposited, the most recent first and the oldest, last (Fig. 1).

Using a chemical process that digests the sediments and leaves the pollen grains intact, microscope slides can be made of the pollen, each slide representing the vegetation of the plant community that existed at a particular point in time.

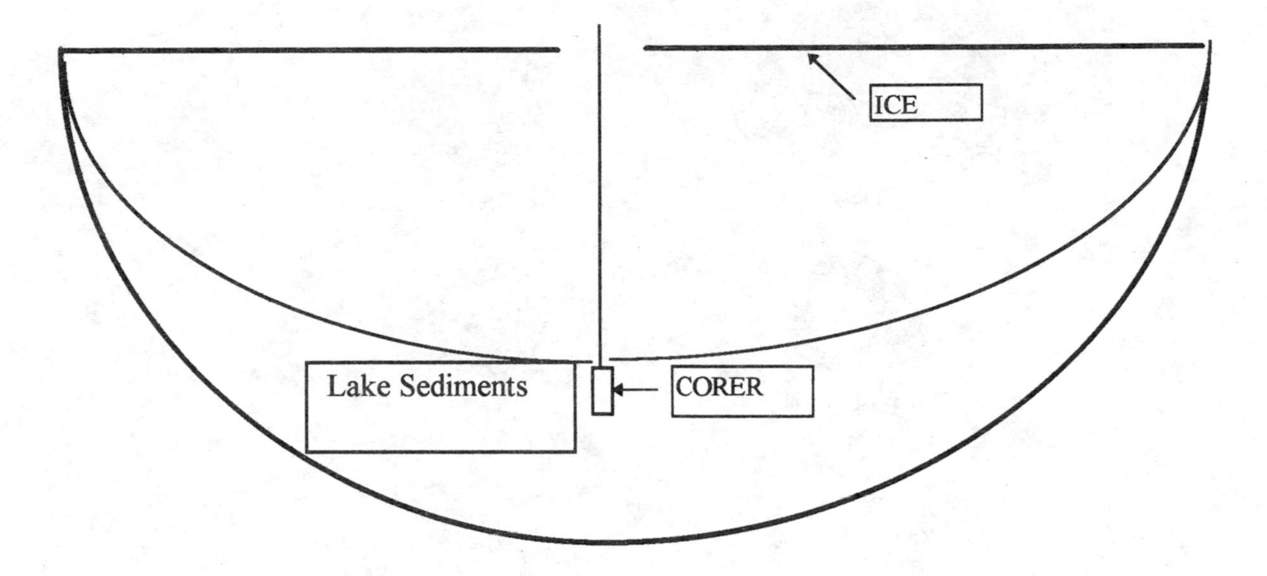

Figure 1. Coring a lake, removing the bottom sediments.

Since pollen grains of different plant taxa look different (Fig. 2.), the plants living at that point in time can be identified (not to species but to genus).

If the plant community can be ascertained, then using the Principle of Uniformity, the climatic conditions at that time can be **inferred**. For example, if at a particular depth in the core, the major pollen types are spruce and fir trees, then it can be inferred that the climate must have been the same, or at least very similar to the boreal forest that exists today in the more northern latitudes (cold and fairly wet). As the major pollen types change at different levels, it can be inferred that the climate must have changed to allow these plants to grow. We will shortly see that this may not be exactly the case, but for the most part, the vegetation responds to changes in climatic conditions.

There are techniques to date these sediments-tell how old they are. The technique used is called **radiocarbon dating**. So **where** a major shift in the pollen types is seen in the core, **when** these changes occurred can be ascertained.

Simply, radiocarbon dating is based on the fact that radioactive carbon, which is presumed to have a more or less constant composition in the earth, both in the past and present, decays at a known rate of 530 years. By measuring the proportion of radioactive carbon in living tissue to that of the nonliving organic material in the sediments, and knowing the rate of decay (the half-life), the age of the organic material can be estimated. It is, in fact, an estimation with the possibility of error. But it is a good estimation nonetheless.

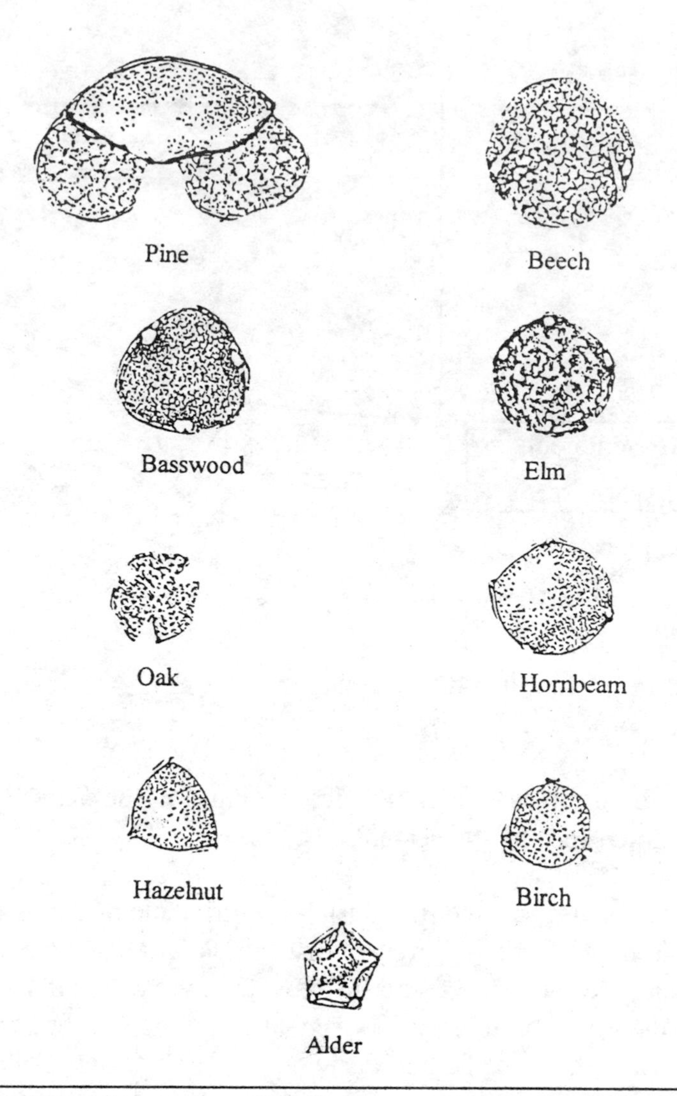

Figure 2. Pollen grains of different taxa look very different. They differ in size (measured in micrometers), shape, and surface characteristics; some have pores, some furrows (slits), some pores and furrows. They can be distinguished relatively easily from one another.

On page 33 is a pollen diagram from Pine Bog in High Point State Park, Sussex County, New Jersey. Note that the pollen of each taxon is measured as a percentage of the entire assemblage at a particular point in the core. Along the "y" axis is the depth of the core as well as the radiocarbon dates, some of which have been extrapolated. By knowing the general composition of vegetation at a particular depth (and time), we can infer the climate that must have been present to sustain that vegetational complex. This pollen diagram comes from the following paper:

Niering, W.A. 1953. The past and present vegetation of High Point State Park, New Jersey. Ecological Monographs 23: 127-148.

This pollen diagram was downloaded from the Internet. Later, data from additional paleoecological studies will be supplied for the construction of other pollen diagrams. The program that incorporates all of these data that can also be downloaded from the Internet is: **SiteSeer** version 1.1 for Windows. The program was designed and written by **John Keltner,** at the Illinois State Museum. The Internet address for accessing this information and additional data is at the National Oceanic and Aeronautics Administration (NOAA) Paleoclimatology Laboratory:

http://www.ngdc.noaa.gov/paleo/paleo.html

Let's look at the diagram:

The oldest date is 13,000 years B.P (older dates and vegetation have been extrapolated). At that point in the core, the dominant pollen is spruce (*Picea*) and pine (*Pinus*). We can infer that the climate was much cooler than it is today because these two taxa are conifers and indicative of the boreal forest. By 9,000 years B.P., the spruce has dropped dramatically and pine is beginning to decline. Deciduous trees are becoming more important with the rise of oak (*Quercus*). Maple (*Acer*) appears as well as birch (*Betula*), willow (*Salix*) and hemlock (*Tsuga*), a conifer. The increase in these species may indicate a climate that was warmer than the earlier dates. By approximately 5,000 years B.P., the dominants are oak, hickory (*Carya*), and beech (*Fagus*) possibly due to the climate becoming drier. At approximately 3,000 years B.P., American chestnut (*Castanea*) becomes prominent. The forest is dominated by oak, chestnut, birch and hemlock, essentially what our forests look like today; this change may indicate that the climate may have gotten cooler and slightly wetter. Note that toward the present, chestnut falls out. Why might this have happened?

Look at alder (*Alnus*) in the pollen record. Alder is a shrub that produces an enormous amount of pollen. It is represented in the diagram in significant percentages from about 9000 years B.P. to the present. Pine too, produces extremely large amounts of pollen. These taxa illustrate one of the short comings of climatic interpretations based upon pollen percentages: Abundant pollen producers will be over-represented in a core, as pine and alder are, and taxa producing small amounts of pollen will be under-represented (insect pollinated species usually do not produce large amounts of pollen). These two taxa probably were much less abundant than the record indicates. Using the above diagram, other limitations of paleoecological reconstructions can be understood. A pollen diagram usually gives a regional interpretation of vegetation and climate since pollen can travel

Pine Bog

(cm) (yrBP) *Acer* *Alnus* *Betula* *Carya* *Castanea* *Corylus* *Fagus* *Fraxinus* *Ostrya-type* *Picea* *Pinus* *Quercus* *Salix* *Tsuga* *Ulmus* *Other trees and sh*

hundreds and sometimes, thousands of miles. It is possible to have pollen grains in the core that do not represent the vegetation present at the site.

Furthermore, change in the pollen percentages may not be due to climate change at all but to some other factor such as an insect infestation of a particular species which wipes it out.

So pollen diagrams give an approximation of the presence and to some degree, abundance of plant species but there can be many different interpretations of the results. The influences of climate change are also subject to interpretation.

Observe the analog vegetation maps for eastern North American. These maps show how vegetation that is present today, was distributed in eastern North American in the past. These maps are based on almost 12,000 fossil pollen samples and 1800 modern pollen samples. Twenty-one pollen taxa were used in the analysis. They are:

Alder	Fir	Pine
Ash	Hawthorn	Prairie grasses
Aspen	Hazelnut	Sedges
Basswood (Linden)	Hemlock	Spruce
Beech	Hickory	Sweet Gum
Birch	Ironwood/Beaked Hazelnut	Sycamore
Elm	Oak	Walnut

The map was constructed from data from the following paper:

Overpeck, J.T., R.S. Webb, and T. Webb III. 1992. Mapping eastern North American vegetation change over the past 18,0000 years: no-analogs and the future. Geology 20: 1071-1074.

Answer the following questions based on the analog map of eastern North American vegetation:

1. What is meant by the phrase "no modern analog?"

2. Briefly explain how the vegetation in the New Jersey area has changed over time (explain the series of maps).

Showtime is a computer animation program. Turn on the computer, start Windows by typing "Win" at the "C" prompt (C>). When the "Program Manager" screen comes on, double click the Showtime icon.

This program shows the postglacial (after the recession - melting, of the Wisconsin glacier) migration of various plant taxa over the last 15,000 years. The controls work like that of a tape player. The program can be paused, started, stopped, etc. Just pick a taxon and click the "run" button.

You will initially see an outline of the Laurentide Ice Sheet superimposed over a map of North America as it appeared approximately 15,000 years B.P. Note as the animation continues, the plants migrate northward as the ice sheet shrinks. Basically, the larger the diameter of the circles that appear, the greater the pollen percentages of that taxon.

The key to the taxa (Genera) are the following:

Quercus - Oak
Fraxinus - Ash
Betula - Birch
Tsuga - Hemlock
Cyperaceae - Sedges (grass-like plants)
Asteraceae - Aster family including ragweed
Fagus - Beech
Castanea - American Chestnut
Populus - Poplar
Liriodendron - Tulip Tree or Yellow Poplar
Liquidambar - Sweet Gum
Nyssa - Black Gum
Ericaceae - Health family (shrubs, e.g. blueberry, rhododendron)

Choose six of the above taxa and answer the following questions about each of them. Stop the simulation using the pause button to gather information to answer these questions.

1. When did each taxon appear in the pollen record and approximately where did they appear in the present United States?

2. When did each taxon reach its maximum percentage in the pollen record (when did the largest circle appear?)?

3. When did the Laurentide Ice Sheet disappear from the continental United States and where was it last located?

4. What could be the explanation for the Asteraceae increasing (remember that ragweed is in this family) around 1,600 A.D.?

Showtime was downloaded from the Internet and is the work of:

John Keltner
Illinois State Museum
1920 South 10 ½ Street
Springfield, Il 62703

The last exercise, as all of the data you have observed, was downloaded from the Internet. You will be asked to retrieve the data for this last exercise or be given a hard copy of it. Each student will be assigned a different site at which pollen was collected as part of a paleoecological study. From these data construct a pollen diagram on the graph paper that is included at the end of this exercise, similar to the Pine Bog diagram, as observed earlier. While the format of the Pine Bog diagram should be followed, construct the diagram following these instructions:

First, graph data at 1000 year intervals or as close to that time scale as possible. Construct the diagram **lengthwise** (sometimes referred to as the "landscape" orientation) on the graph paper. Depth of the core and the appropriate radiocarbon dates should be included on the "y" axis of the diagram. Let each box represent 250 years going down. List the various taxa on the top of the diagram, along the "x" axis. Each taxon will need a number of squares to represent its percentage of the total pollen sample, at a specific depth. Going across, for each taxon, let each square on the graph paper represent 5% of the sample, e.g. if oak, at a particular depth, comprises 50% of the total pollen assemblage, shade in 10 boxes to represent the percentage of the total assemblage that oak represents at that particular depth. **Review the data before the diagram is constructed and plan how many boxes each taxon will require**. When all the percentages, at all depths for a particular taxon are plotted, a pattern for the abundance, decline or increase, for that taxon can be seen through time. Do this for all 16 entries. Use the Pine Bog diagram as a guide.

There is only one question concerning your pollen diagram: Interpret the diagram, inferring the climate change associated with the change in the percentages of the important taxa that you have graphed and based upon your knowledge of present North American biomes. What other factors could have resulted in the pollen assemblages that you have graphed?

Name___________________________

EXERCISE 5: ECOLOGICAL MODIFICATIONS IN PLANTS AND ANIMALS

Organisms that are successful in their environments have developed adaptations that allow them to survive through the entire ranges of variation for many environmental factors. These organisms have a wide **tolerance limit** for these factors and thus can withstand the stresses of their environment.

To understand why plants and animals are successful in their environments, we would look at organisms living in extreme environments and see what adaptations they have developed. Later in this exercise, you will be asked to consider some of these environments and how certain organisms cope.

The first part of this exercise deals with modifications in leaf morphology (shape) for surviving in different environments.

Water is a basic need for all living organisms, let alone plants. But plants can be found in environments ranging from little water to living in water itself and everything in between. Plants that live in water deficient environments are called xerophytes (xeric environment), plants living in aquatic environments are called hydrophytes (hydric environment) and those that live where water is in optimum amounts or where there is a moderate amount of water are mesophytes (mesic environment). These plants have leaves that are adapted to surviving in these conditions and as such, are different in structure, if not function. Before we look at these modifications, we need information on basic leaf structure and function.

The major function of leaves is that they are the site of PHOTOSYNTHESIS--THE UTILIZATION OF SUNLIGHT AS AN ENERGY SOURCE TO PRODUCE FOOD (COMPLEX ORGANIC MOLECULES) WITHIN THE PLANT!!! Plants produce their own food (that is why they are referred to, in ecological terms, as producers or autotrophs). A secondary function of leaves is that excess water is removed from the plant through the leaves; a process called transpiration.

Leaves come in all shapes and sizes. They absorb the necessary gases (chiefly carbon dioxide) for the photosynthetic process and expel oxygen and water vapor. Before we look at the modifications of leaves in different habitats, lets look at the anatomy of a "typical leaf," in terms of its structure and function. Please refer to the diagram on the following page.

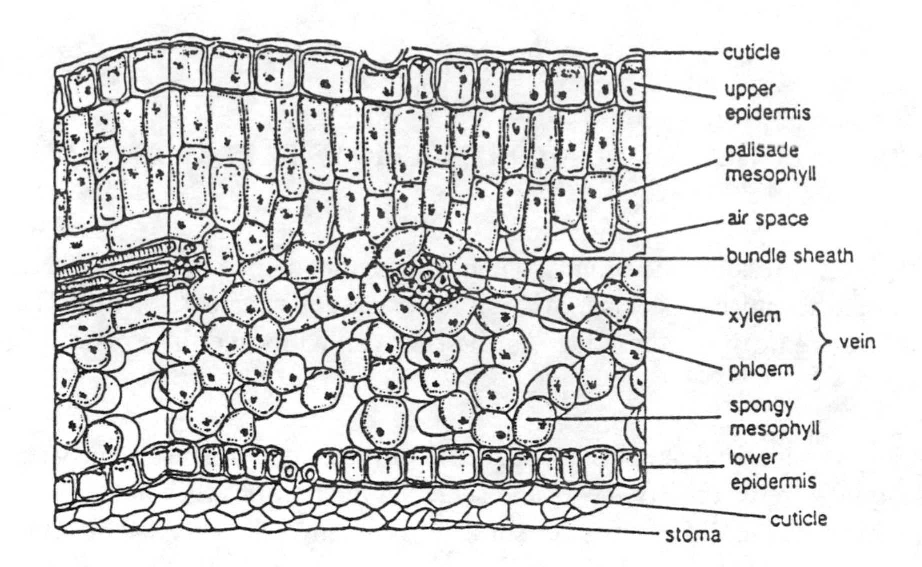

The **epidermis** is a single layer of cells in thickness and is continuous over the top and bottom of the leaf. The epidermis is usually covered with a waxy cuticle. **Guard cells** are components of the epidermis and occur in pairs. They are separated by a pore-the **stoma** (plural = **stomata**). The stomata allow gas exchange between the leaf and the environment, CO_2 in and O_2 and water vapor out. The size of the stoma is controlled by the guard cells which can shrink and swell.

The **mesophyll** is the tissue between the upper and lower epidermis. It is photosynthetic in function because these cells contain the vast majority of **chloroplasts** (the sites of photosynthesis). The mesophyll is further divided by the two cell types which comprise it - the palisade mesophyll and the spongy mesophyll. The palisade layer is closest to the upper epidermis and the spongy layer to the lower epidermis. The spongy mesophyll contains many air spaces and is a loosely arranged tissue.

Also contained in a leaf is the **vascular** (conducting) tissue. Made up of veins, this tissue carries materials to and from the leaf.

In this laboratory exercise the student will first observe a typical, mesophytic leaf in cross-section and then observe a number of leaves from different habitats.

Generally, hydrophytic leaves contain many air chambers. Since they are either submerged or floating, there is a reduction in tissues that support the plant and generally have cells with thin walls. For floating leaves, the stomata are only on the upper epidermis. Finally, because they are surrounded by a liquid medium which bathes them in minerals and nutrients, hydrophytes have no need for a sophisticated vascular system nor do they have to conserve water. As a result, the epidermis has a thin cuticle.

Xerophytic plants have the opposite problems, they must conserve water, prevent water loss. Generally, these plants have stomata located below the epidermis (recessed from the epidermis). This will allow gas exchange but minimize the loss of water through transpiration. The stomata are usually closed at night when photosynthesis is not occurring, to prevent water loss.

Xerophytes have a thick, waxy cuticle made of **cutin** which is a water-proofing. There is also substantial supportive tissue which not only supports the leaf but also prevents water loss. Many plants in xeric environments exhibit leaf rolling. The leaf actually rolls up into a tight cylinder. This behavior isolates many of the stomata on the upper epidermis and buffers them from the environment. While we consider xeric habitats to be mostly dry, desert-like environments, there is also "physiological drought" where due to the chemical and physical nature of the environment, water is not readily available to the plants. Finally, xerophytes have large amounts of water-storage tissue. These plants are fleshy and sometimes referred to as succulents.

Observe the cross section of the lilac (*Syringa*). This plant is mesic in terms of its relationship to moisture and represents a "typical leaf." Locate all the structures that are found in the diagram of the leaf.

Obtain a slide of pondweed (*Potamogeton*). This is a hydrophyte. Draw and label this leaf. Compare it to the lilac. List the differences and similarities.

Potamogeton:

Differences from the lilac:

Similarities:

Compare the *Yucca* leaf, a xerophyte, to the lilac. Draw and label this cross-section. List the differences and similarities.

Yucca:

Differences from the lilac:

Similarities:

Now obtain a slide with the three different leaf types: Hydrophyte, Mesophyte and Xerophyte. Draw and label all three.

Look at the slides of the five "unknowns" and fill in the chart as to their characteristics. In the last column, state the reasons as to why you classified them as hydrophytic, mesophytic or xerophytic.

SPECIES	Presence and Size of Air Chambers (many or few, large or small)	Amount of Supporting Tissue (much or little)	Leaf Modifications: (Stomata, Cuticle, etc.) (Many or few, thick or thin)	Vascular Tissue (much or little)	Habitat: Xeric, Mesic or Hydric (and reason)
Typha					
Pinus					
Zea					
Ammophila					
Myriophyllum					

Answer the following questions:

1. Submerged plants do not need large amounts of supporting tissue. Why?

2. Summarize the differences between hydrophytes, mesophytes and xerophytes.

Observing the ecological modifications of animals is more difficult than plants. Many times individuals of the same species will show variations because they live in different environments.

Bergmann's Rule: Members of a species possessing smaller body-size are found in the warmer parts of their range, larger members are found in cooler parts of their range. Not only does this appear true for homeotherms (organisms that can regulate their body temperature) but for poikilotherms (organisms that cannot regulate their body temperature) as well.

Allen's Rule: Extremities of animals (tails, ears, bills) are shorter on individuals in the cooler parts of their range than the warmer parts.

These two rules are generally true and are based on the fact that large body-size and short appendages and extremities give less surface area per volume of body and thus minimize heat loss from the body in colder climates.

Answer the following questions:

1. The jackrabbit (*Lepus townsendii*) of the southwestern United States has extremely large ears which are highly vascularized (contain many blood vessels). Why?

2. What behavioral adaptation have animals that live in hot, dry climates developed to cope with the rigors of their environment?

3. What physiological adaptation have animals in cold climates developed to cope with their environment?

4. Where in the continental United States would you expect to find the smallest members of the white-tailed deer (*Odocoileus virginianus*) population? Where would the largest members be found?

5. The timber wolf (*Canis lupus*) was once found throughout much of North America (except in parts of the southeastern United States). Where were the smallest members of that population found? The largest members?

Name________________________

EXERCISE 6: ECOLOGICAL SAMPLING: A FOREST ECOSYSTEM

For many ecologists, **sampling** of various environmental parameters or units of study is the major emphasis of a research project. Many times it is impossible to look at every individual in a population or community. So a random estimation (**sample**) of the nature and composition of a particular area or a random estimation of the number of organisms is undertaken. A sampling regime is established. Sampling is in the realm of statistics, statistical sampling! There is always error when sampling is undertaken and it is hoped that the techniques used for the study minimize this error. As the science of ecology became more sophisticated, better and more accurate methods of sampling were designed. This exercise and the following one utilize common statistical sampling methods for forest and meadow ecosystems, respectively. For these exercises, the forest and the meadow have been brought to the classroom. These methods illustrate the strengths and weaknesses of ecological sampling.

While it would be nice to have a laboratory class go out into the forest and actually have a field experience, the logistics of such an exercise sometimes make it impossible. Some students would get lost as soon as they left the concrete and others would panic because there would be no malls nearby.

In this exercise, each pair of students will first work with a "forest" on either a north-facing slope of a mountain or the south-facing slope. When all the data have been gathered from the first slope, get a "forest table" for the other slope. Each pair of students will gather data from both the north-facing and the south-facing slopes.

Observe the figure below: What factors determine the distribution and abundance of vegetation (in this case, trees) on the north and south-facing slopes? Consider the microhabitat differences between the two slopes. What other factors are influenced by differences in incident solar radiation?

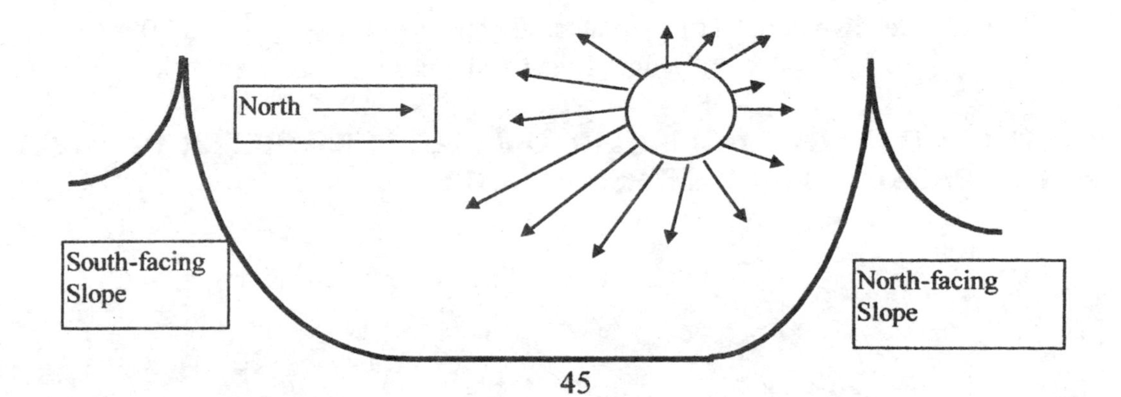

PROCEDURE:

A technique called the Point-Quarter method will be employed to sample the forest. On the forest table, run a meter stick along one side of the length of the table. This is a **transect** line, simply a straight line run along a compass point at which data will be gathered. At 20 centimeter intervals, place a ruler perpendicular to the transect. This is a "point" and data will be gathered at each point. The point has been divided into four quarters (the ruler perpendicular to the transect line). Record the tree species (consult the color chart for the species of each tree) by indicating its diameter at breast high (dbh.) value in Table 1 - the data sheet for either the south face or north face, which ever is being used first. **There will be only 4 dbh's per point (only 4 in each column).** Move 20 centimeters and record the data for the next point and so on. Collect data for 20 points for each slope. Refer to the following diagram.:

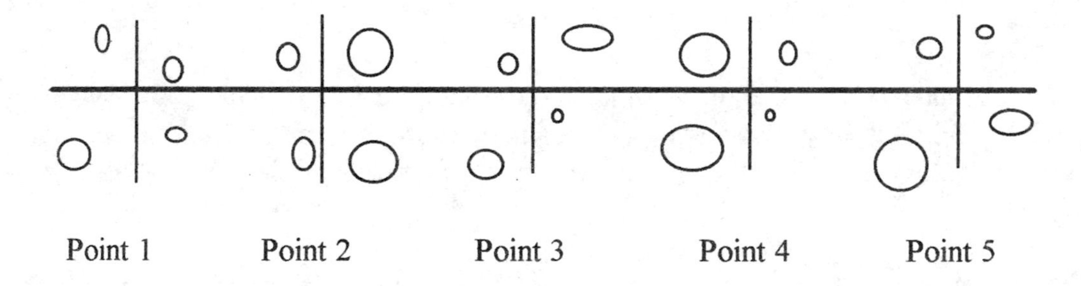

Point 1 Point 2 Point 3 Point 4 Point 5

Calculate the following for each tree species and fill in the data sheets for the south face (Table 2) and north face (Table 3), respectively:

a. relative density $= \dfrac{\text{\# of individuals of one species}}{\text{sum of the individuals of all species}} \times 100\%$

b. relative dominance $= \dfrac{\text{total basal area of one species}}{\text{sum of the basal areas of all species}} \times 100\%$

TO DO THIS, CONVERT EACH DBH INTO BASAL AREA. REFER TO TABLE 3: DBH TO BASAL AREA CONVERSION TABLE

c. frequency = $\frac{\text{total } \textbf{points} \text{ at which a species occurred}}{\text{total number of } \textbf{points} \text{ sampled}} \times 100\%$

d. relative frequency = $\frac{\text{frequency of } \textbf{one} \text{ species}}{\text{sum of frequencies of } \textbf{all} \text{ species}} \times 100\%$

e. importance value = relative frequency + relative dominance + relative density

IV = RF + RB + RD

Basic information in ecological sampling is the number of individuals in a specific area. This is called the ABUNDANCE (N). The number of individuals per unit area is the DENSITY (D). When comparing the density of one species to all species, the RELATIVE DENSITY (RD) becomes significant. The relative density is the proportion of the total number of organisms of one species compared to the total number of all species in the area (divide the density of one species by the total densities of the other species).

The FREQUENCY (F) of an organism is the number of times it appears in the sample. In the case of the point-quarter method it is the number of sampling **points** a species appears at divided by the total number of sampling **points**. The RELATIVE FREQUENCY (RF) is the frequency of one species divided by the sum of all frequencies of all species.

The DOMINANCE (D) of a species, in the context of this exercise, is the coverage or the ground occupied by a species. It is expressed as the area covered by a species and is determined, for this exercise, by measuring the **diameter** of each tree of each species, converting the diameter (taken breast height in the field) to an **area**. The RELATIVE DOMINANCE (RB) is the proportion of the dominance of one species to the dominance of all species. It is calculated by first converting dbh for each individual of a species, finding the sum, the **total area for each species**, and divided by the total sum of areas for **all** species.

Finally, the IMPORTANCE VALUE (IV) is the sum of the RELATIVE FREQUENCY (RF) + RELATIVE DENSITY (RD) + RELATIVE DOMINANCE (RB). It will therefore range from) 0 - 300%. It gives an overall estimate of the "influence" or "importance" of each species in the community. It reveals the "nature" of each species in the community.

Once all of the data sheets are completed, answer the following questions:

1. How do the two slopes differ in species composition of their forests?

2. List the two species that had the highest frequencies in each forest?

3. List the two species that are the most dense in each forest?

4. Which four species are the most important in each forest?

5. What factors are responsible for variations in the forest compositions. What is the limiting factor in the forests? Give a detailed explanation.

TABLE 1. DATA SHEET FOR TREE SPECIES AND DBH - SOUTH FACING SLOPE

TRANSECT POINTS

SPECIES	1	2	3	4	5	6	7	8	9	10	11	12	13	14	15	16	17	18	19	20
Sugar Maple Black Square																				
Scarlet Oak Red Square																				
Red Oak Red Circle																				
American Beech Black Circle																				
Chestnut Oak Black Octagon																				
Red Maple Red Octagon																				
Flowering Dogwood Green Octagon																				
White Ash Red Parallelogram																				
White Oak Green Circle																				
Shagbark Hickory Green Square																				

TABLE 1'. DATA SHEET FOR TREE SPECIES AND DBH - NORTH FACING SLOPE

TRANSECT POINTS

SPECIES	1	2	3	4	5	6	7	8	9	10	11	12	13	14	15	16	17	18	19	20
Sugar Maple Green Circle																				
Black Birch Black Octagon																				
Red Oak Red Octagon																				
American Beech Green Octagon																				
Chestnut Oak Black Square																				
Red Maple Red Circle																				
Yellow Birch Green Square																				
White Ash Red Square																				
White Oak Black Circle																				
Tulip (Yellow) Poplar Red Parallelogram																				

TABLE 2. DATA SHEET FOR FOREST SAMPLING - POINT QUARTER METHOD

SOUTH-FACING SLOPE

SPECIES	P = # of Points	Q = # of Trees	Basal Area	Frequency	RF = Relative Frequency	RB = Relative Dominance	RD = Relative Density	IV = Importance Value
Sugar Maple								
Scarlet Oak								
Red Oak								
American Beech								
Chestnut Oak								
Red Maple								
Flowering Dogwood								
White Ash								
White Oak								
Shagbark Hickory								
SUMS		=N			=100%	=100%	=100%	=300%

TABLE 2'. DATA SHEET FOR FOREST SAMPLING - POINT QUARTER METHOD

NORTH-FACING SLOPE

SPECIES	P = # of Points	Q = # of Trees	Basal Area	Frequency	RF = Relative Frequency	RB = Relative Dominance	RD = Relative Density	IV = Importance Value
Sugar Maple								
Black Birch								
Red Oak								
American Beech								
Chestnut Oak								
Red Maple								
Yellow Birch								
White Ash								
White Oak								
Tulip (Yellow) Poplar								
SUMS		=N			=100%	=100%	=100%	=300%

TABLE 3. DIAMETER BREAST HIGH (DBH) TO BASAL AREA CONVERSION CHART.

DBH	BASAL AREA
4.00	0.0012
4.50	0.0015
5.00	0.0019
5.50	0.0023
6.00	0.0028
6.50	0.0033
7.00	0.0038
7.50	0.0044
8.00	0.0050
8.50	0.0056
9.00	0.0063
9.50	0.0070
10.00	0.0078
10.50	0.0086
11.00	0.0095
11.50	0.0103
12.00	0.0113
12.50	0.0122
13.00	0.0133
13.50	0.0143
14.00	0.0154
14.50	0.0165
15.00	0.0177
15.50	0.0189
16.00	0.0201
16.50	0.0214
17.00	0.0227
17.50	0.0241
18.00	0.0254
18.50	0.0269
19.00	0.0284
19.50	0.0299
20.00	0.0314

Name________________________

EXERCISE 7: ECOLOGICAL SAMPLING - A MEADOW ECOSYSTEM

In the previous exercise, we sampled, albeit vicariously, a forest ecosystem using the point-quarter method along a transect. We now turn our attention to a different ecosystem which necessitates a different kind of sampling regime and one that can be more random.

A meadow is an early succesional stage in a terrestrial ecosystem. Picture an abandon farm field. In the first year after it is allowed to lie fallow, grasses become abundant and dominate this area. But by the third or fourth year, the grasses have essentially been replaced by a variety of herbaceous species (weeds to the uninformed!).

Here in the Northeast, it is quite beautiful to observe the yellows, whites, reds, purples and greens of a meadow in the late summer and early fall..

In this exercise, the class will work with a "meadow" that has been produced artificially in a similar manner as the forest in the previous exercise. However, instead of running a transect (a straight line) and dividing each point along the transect into quarters, two new methods will be employed using a "plot" or quadrat. Plots can be a variety of geometric shapes which can be randomly established within a study area or established along a transect. Quadrats are plots that are usually square or rectangular.

PROCEDURE

Once a quadrat is established within a study area, all the plants are counted and identified within the plot. Whether randomly established or along a transect or even a grid, the more quadrats established, the more accurate the sampling will be (makes sense!!).

Each pair of students will establish a transect on the "meadow table" and count and identify the vegetation within 20 transparent quadrat overlays along each transect. Establish a second transect if twenty plots cannot be made along the first. The overlay is 6"^{2} and while rectangular plots seem to be more accurate for vegetation analyses, these will do fine. Run each quadrat 1" from the previous one along transect(s). Record the number of each species in Table I. Then repeat the exercise randomly distributing 20 quadrats over the entire "meadow table" (just flip the transparency onto the meadow table, **randomly**!). Record these data (by the way, data is a plural term; datum is singular) in Table 2.

For plants that are partially in the quadrat, if the plant is more than half in the quadrat count it, if not, ignore it.

The key to the plants on the meadow table is the following:

SYMBOL	COMMON NAME	SCIENTIFIC NAME
1. blue circle	Spotted Knapweed	*Centaurea maculosa*
2. orange circle	Rough-Stemmed Goldenrod	*Solidago rugosa*
3. yellow circle	Black-Eyed Susan	*Rudbeckia hirta*
4. blue star	Chicory	*Chichorium intybus*
5. green circle	Pearly Everlasting	*Anaphalis margaritacea*
6. black star	Daisy Fleabane	*Erigeron annuus*
7. white square or circle	Heal-All	*Prunella vulgaris*
8. red circle	Queen Anne's Lace	*Daucus carota*
9. green star	Evening Primrose	*Oenothera biennis*
10. red star	Common Mullein	*Verbascum thapsus*

These species, along with many, many others, including a number of grasses are found in meadow ecosystems. When the calculations are made in this exercise, they may not necessarily reflect the actual abundance and frequency of these species in a real meadow. This is just a simulation.

Calculate the following for each species and fill in Tables 3 and 4:

a. $\text{percent frequency} = \dfrac{\text{total \# of samples or quadrats containing a species}}{\text{total \# samples or quadrats}} \times 100\%$

The percent frequency means: out of all the quadrats (in this case, 20), how many quadrats was a species found in?? This is done for each species. Count the number of quadrats a species was present in and divide by 20!!

b. average density = number of plants per quadrat

Average density means: the average number of individuals of a species found in one quadrat. Add all of the individuals of one species found in all 20 quadrats and divide by 20 - an average!!!

After all the data sheets are completed, answer the following questions:

1. How would the accuracy of the results (density and frequency) be affected if the number of quadrats were increased. What if the size of the quadrats were made larger?

2. How did the results differ between the random method of establishing quadrats and running them along a transect?

3. There are three patterns of distribution for organisms, uniformly, random, and clumped. How is each species in this experiment distributed and give reasons why they might be distributed in that manner.

TABLE 1. SPECIES COUNTS WITHIN QUADRATS ESTABLISHED ALONG TRANSECTS.

	QUADRATS																			
SPECIES	1	2	3	4	5	6	7	8	9	10	11	12	13	14	15	16	17	18	19	20
SPOTTED KNAPWEED																				
ROUGH-STEM GOLDENROD																				
BLACK-EYED SUSAN																				
CHICORY																				
PEARLY EVERLASTING																				
DAISY FLEABANE																				
HEAL-ALL																				
QUEEN ANNE'S LACE																				
EVENING PRIMROSE																				
COMMON MULLEIN																				

TABLE 2. SPECIES COUNTS WITHIN RANDOMLY ESTABLISHED QUADRATS.

	QUADRATS																			
SPECIES	1	2	3	4	5	6	7	8	9	10	11	12	13	14	15	16	17	18	19	20
SPOTTED KNAPWEED																				
ROUGH-STEM GOLDENROD																				
BLACK-EYED SUSAN																				
CHICORY																				
PEARLY EVERLASTING																				
DAISY FLEABANE																				
HEAL-ALL																				
QUEEN ANNE'S LACE																				
EVENING PRIMROSE																				
COMMON MULLEIN																				

TABLE 3. PERCENT FREQUENCY AND AVERAGE DENSITY OF SPECIES WITHIN QUADRATS ESTABLISHED ALONG TRANSECTS.

SPECIES	PERCENT FREQUENCY	AVERAGE DENSITY
SPOTTED KNAPWEED		
ROUGH-STEM GOLDENROD		
BLACK-EYED SUSAN		
CHICORY		
PEARLY EVERLASTING		
DAISY FLEABANE		
HEAL-ALL		
QUEEN ANNE'S LACE		
EVENING PRIMROSE		
COMMON MULLEIN		

TABLE 4. PERCENT FREQUENCY AND AVERAGE DENSITY OF SPECIES WITHIN RANDOM QUADRATS.

SPECIES	PERCENT FREQUENCY	AVERAGE DENSITY
SPOTTED KNAPWEED		
ROUGH-STEM GOLDENROD		
BLACK-EYED SUSAN		
CHICORY		
PEARLY EVERLASTING		
DAISY FLEABANE		
HEAL-ALL		
QUEEN ANNE'S LACE		
EVENING PRIMROSE		
COMMON MULLEIN		

Name____________________

EXERCISE 8: PREDATOR-PREY RELATIONSHIPS

One of the classic examples in ecology of a sigmoid (S-shaped) growth curve; the fluctuations or oscillations of animal populations, is the abundance of the Canada lynx and snowshoe hare of the boreal forest biome. Since North America has been settled, the Hudson's Bay Company of Canada has kept records of the pelts of the fur-bearing creatures trapped each year. These data indirectly reflect the changes in the populations of each organism annually.

Essentially, the snowshoe hare has a cycle between 8 and 11 years where the population will increase to great abundance and then decline. The lynx population has a cycle between 8 and 12 years and basically follows the snowshoe hare population fluctuations with approximately a one year lag. Since the snowshoe hare is one of the primary foods for the lynx, it would be assumed that their population cycles would be intimately connected. While this is true, and the objective of this laboratory is to show this relationship, there are other factors involved in the rise and fall of the two populations. However, it is beyond the scope of this course to deal with them. This laboratory is designed to show the relationships between a predator species and its prey.

For this exercise, the student is asked to make the following assumptions:

1. The surviving number of hares will double in the next run of the experiment (test run). In other words, those hares that aren't eaten by the lynx will double their numbers in the next test run.

2. In each test run, at least 10 hares are initially present in the forest (by immigration if necessary). So if the number of hares drops below 5 after a test run, bring the number of beans (hares) back up to 10. There should never be less than 10 beans in the dish as a test run is begun.

3. The carrying capacity (the maximum number of hares that can be supported) of this boreal forest ecosystem is 100. There should never be more than 100 beans in the dish during any test run.

4. In each test run at least one lynx is initially present in the forest (if they all die or emigrate out, at least one immigrates into the forest).

5. For a single lynx to survive in the forest, it must capture at least 5 hares (if it doesn't capture at least 5 hares, it will either starve or emigrate out of the forest to search for food).

6. For every 5 hares that a lynx captures, it will produce 1 offspring (if one lynx captures 12 hares it will produce 2 offspring, if 7 hares are captured, 1 offspring, if 3 hares are captured, no offspring). If a second lynx captures just 4, it will produce no offspring even though a first lynx may have caught 8 (leaving 1 offspring).

PROCEDURE

Working in pairs, each team should obtain a spoon, dish and 100 beans. Together, with the instructor, fill in the first 4 test runs on the Data Sheet. As one student performs the experiment, have the other record the results. A total of **30** test runs will be made so after the first four, done with the instructor, run the experiment 26 more times.

Test Run 1

Begin with 10 hares and 1 lynx in the forest (place 10 beans in the dish). The beans are the hares, the spoon is the lynx and the dish is the boreal forest. The capture of the hares by the lynx is simulated by scooping the spoon once through the dish for each lynx present. When this is done, the student should not consciously try to scoop up beans; don't look at the dish or use your other thumb to scoop up the beans. For test run 1, assume that the lynx catches no hares. In Table I enter 0 for prey captured by predator 1 of the first test run. Also enter 0 for offspring produced by the predator. Then fill in test run 1 of Table II. The 10 surviving hares double their numbers (first assumption-add 10 more beans to your dish) and since the lynx either starved or emigrated, another lynx must immigrate into the forest.

Test Run 2

Again, as there were no predator survivors in the first test run, assume that another lynx immigrates into the forest. Scoop once through the dish, picking up 3 beans. In Table I enter 3 for prey captured by predator 1 of the second test run, and 0 for predator offspring. Then fill in test run 2 of Table II. Seventeen hares survive; therefore, add 17 more beans to the dish.

Test Run 3

Since there were no offspring, and the only lynx couldn't survive, another lynx must have entered the boreal forest. Scoop once through the dish, this time picking up 8 beans. In Table I enter 8 for the prey captured by predator 1 of the third test run **and** enter 1 for the offspring produced by this predator (remember, it caught at least 5 hares!!!). Fill in test run 3 of Table II. Twenty-six hares survived so add 26 more beans to the dish.

Test Run 4

There are 2 predators in test run 4 (the predator that survived and the one predator offspring of test run 3). Now two scoops through the beans will be made. For the first scoop, remove 11 beans. In Table I enter 11 for prey captured by predator 1 of the fourth test run, and 2 for the offspring produced by this predator. For the second scoop, pick up 8 beans. In Table I enter 8 for prey captured by predator 2 of the fourth test run, and 1 for the offspring produced by this predator. Then fill in test run 4 of Table II. Thirty-three hares survive; therefore, add 33 more beans to the dish.

EXERCISE COMPLETION

Complete the exercise by continuing the predation of snowshoe hares by lynx. Continue to make as many passes through the center of the dish (without looking!!!) as the number of lynx present in that particular test run. Do not intentionally scoop up all the beans. In Table II remember that Initial Prey never falls below 10 or more than 100, and Initial Predators never falls below 1. When all the test runs have been completed and the hunt is over, graph the results following the directions on the bottom of Table II.

Your graph should look something like this:

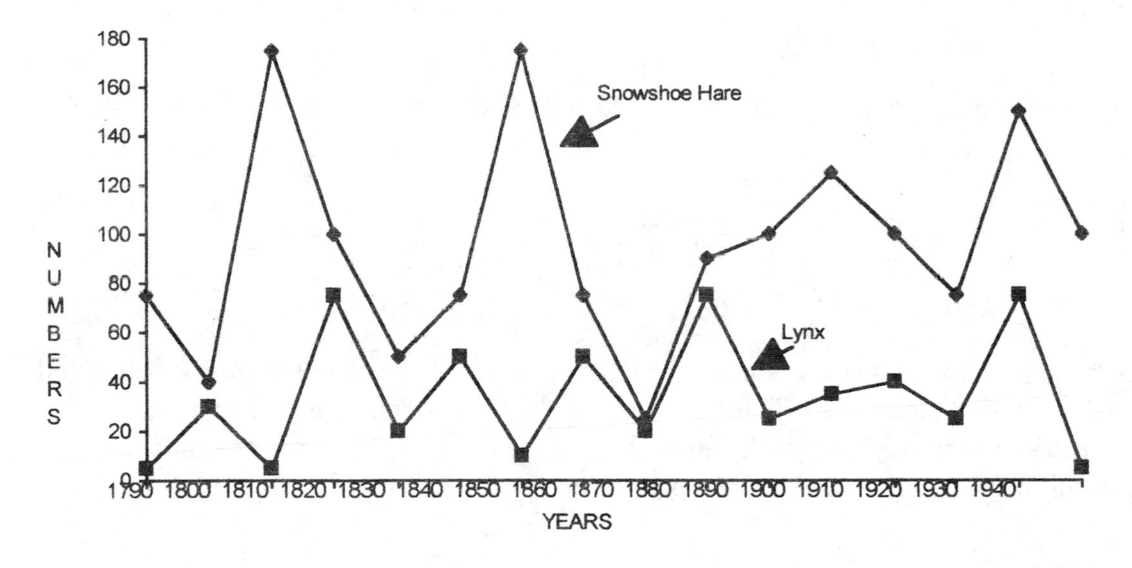

This is the graph based upon the data of MacLulich (1937). See the references at the end of this lab. Although your numbers and time differ, your graph should be similar.

QUESTIONS

1. Which population (predator or prey) shows the first increase in numbers?

2. Does a peak in lynx population occur simultaneously as, or after a peak in the hare population? Why?

3. What parameter apparently determines the size of the lynx population in the forest in any given test run?

4. What parameter apparently causes the decline of the hare population?

5. What other factors might control the populations of the lynx and hares?

6. In real life do you think the hare population would completely disappear in the forest? Why?

7. The three basic patterns of distribution of individuals of a population are: uniformly distributed throughout the habitat, randomly distributed throughout the habitat, and clumped in groups but randomly distributed throughout the habitat. What is the most common pattern and why would populations exhibit this pattern and the other two?

REFERENCES

S.C. Kendeigh. 1974. Ecology: with special reference to animals and man. Prentice-Hall, Inc. Englewood Cliffs, New Jersey. 475 p.

MacLulich, D.A. 1937. Fluctuations in the numbers of the varying hare (*Lepus americanus*). Univ. Toronto Studies, Biol. Ser., No. 43.

Odum, E.P. 1971. Fundamentals of ecology. W.B. Saunders Co. Philadelphia, Pa. 575 p.

WA-TOR

Wa-tor is a shareware program written by Dr. Warren Kovach of Kovach Computing Services in Wales, U.K.

Wa-tor is a very simple computer simulation of the predator-prey relationship between sharks and "fish." In it, the student sets the following parameters for both the fish and sharks: breeding age and initial number. Starvation time (how long the shark can live without eating a fish) can also be set for the sharks. The object of this exercise, for us, is to perpetuate the two populations and produce a sigmoid (S-shaped) growth curve for both populations, and then manipulate the parameters..

Procedure

As in the previous computer labs, turn on the computer, at the "C" prompt, type "Win" which will start Windows. When the program manager screen comes on, double click the Wa-tor icon, which will start the program.

The main Wa-tor screen appears. Click the menu choice "Options," at the top of the screen. You will be able to set the parameters mentioned above. For now though, accept the default choices (the default choices are the ones already on the screen) by clicking "O.K.." At the top of the screen click the "Game" menu choice and then click "Start." The simulation begins. The screen will show the fish increasing when there are low numbers of sharks but as the sharks increase in number, the fish begin to decrease. You can follow the actual change in numbers by the data box to the right. Now click the "Graphs" menu choice at the top of the screen. Of the three choices, click on "Population." A graph will appear on the screen showing the fluctuation in numbers of both the sharks and fish. The graph plots the populations for 100 chronos (the time interval used) and then begins to replace the graph with the data for the next 100 chronos (and will continue for as long as the simulation is run).

Follow the simulation through 200 chronos noting the fluctuations of both populations on the graph. Both populations are exhibiting S-shaped growth curves (more or less). Stop the simulation by going to the "Games" menu choice and click "Stop."

We will now start manipulating the parameters to see the effects on both populations. To assess the effect of changing any one parameter, we must keep all the other parameters constant and manipulate one parameter at a time.

At the "Games" menu choice click "New." Now click "Options" and while keeping all the other parameters the same, change the breeding age of the fish to "4." This is done by moving the cursor (with the mouse) to the left of the number in the breeding age box. Use the delete key and erase the "2" and type in "4." Click "O.K.", click the "Games" menu and then click "Start." Once the simulation starts, click the "Graphs" menu and then click on the "Populations" graph. Follow the simulation for 200 chronos, noting how the populations fluctuate. Then stop the simulation.

What is the effect on both the sharks and fish by increasing the breeding age of the fish??

How does this graph compare to the first?

Begin a new simulation as before. Now change, under the "Options" menu, the breeding age of the sharks to 5 rather than 10. Keep all the other parameters the same as in the first simulation. Make sure you change the breeding age of the fish back to 2. Remember: to make some valid conclusions we can only manipulate one parameter at a time!! Once the simulation begins, click on the "Populations" graph under the menu heading of "Graphs." Run the simulation for at least 200 chronos making the same observations as before, then stop the simulation.

What happens??? Why did it happen??

Start a new game, return to the original parameters: Fish-breeding age = 2, initial number = 200, Sharks-breeding age = 10, initial number = 5, starvation time = 3. Change the breeding age of the sharks to 20. Run the simulation and observe the graph for at least 200 chronos.

What can you conclude about the impact of shark predation on the fish population with these parameters??

Exercise Conclusion:

Design and run four different simulations. List the parameters and then your conclusions after running each simulation for at least 200 chronos. Remember, to get meaningful data, from which you can draw a conclusion, just vary 1 parameter at a time, keeping the others the same as in our initial simulation. You may keep all the fish parameters the same and varying one of the shark's or keep the shark parameters the same and vary the fish.

TABLE 1. DATA COLLECTION SHEET FOR PREDATOR-PREY LAB

RUNS OF THE EXPERIMENT

# OF LYNX:		1	2	3	4	5	6	7	8	9	10	11	12	13	14	15	16	17	18	19	20
1	C																				
	O																				
2	C																				
	O																				
3	C																				
	O																				
4	C																				
	O																				
5	C																				
	O																				
6	C																				
	O																				
7	C																				
	O																				
8	C																				
	O																				
9	C																				
	O																				
10	C																				
	O																				
11	C																				
	O																				
12	C																				
	O																				
13	C																				
	O																				
14	C																				
	O																				

C = HARES CAPTURED BY EACH LYNX; O = # OF OFFSPRING BY EACH LYNX

TABLE 1. DATA COLLECTION SHEET FOR PREDATOR-PREY LAB

RUNS OF THE EXPERIMENT

# OF LYNX:		1	2	3	4	5	6	7	8	9	10	11	12	13	14	15	16	17	18	19	20
15	C																				
	O																				
16	C																				
	O																				
17	C																				
	O																				
18	C																				
	O																				
19	C																				
	O																				
20	C																				
	O																				
21	C																				
	O																				
22	C																				
	O																				
23	C																				
	O																				
24	C																				
	O																				
25	C																				
	O																				
26	C																				
	O																				
27	C																				
	O																				
28	C																				
	O																				

C = HARES CAPTURED BY EACH LYNX; O = # OF OFFSPRING BY EACH LYNX

TABLE 1. DATA COLLECTION SHEET FOR PREDATOR-PREY LAB

RUNS OF THE EXPERIMENT

# OF LYNX:		1	2	3	4	5	6	7	8	9	10	11	12	13	14	15	16	17	18	19	20
29	C																				
	O																				
30	C																				
	O																				
31	C																				
	O																				
32	C																				
	O																				
33	C																				
	O																				
34	C																				
	O																				
35	C																				
	O																				
36	C																				
	O																				
37	C																				
	O																				
38	C																				
	O																				
39	C																				
	O																				
40	C																				
	O																				
41	C																				
	O																				
42	C																				
	O																				

C = HARES CAPTURED BY EACH LYNX; O = # OF OFFSPRING BY EACH LYNX

TABLE IIa

TEST RUN OF THE EXPERIMENT

	1	2	3	4	5	6	7	8	9	10	11	12	13	14	15
INITIAL PREY (HARES)															
TOTAL PREY CAPTURED (HARES)															
PREY THAT SURVIVED (HARES)															
INITIAL PREDATORS (LYNX)															
SURVIVING PREDATORS (LYNX)															
PREDATOR OFFSPRING (LYNX)															

TABLE IIb

TEST RUN OF THE EXPERIMENT

	16	17	18	19	20	21	22	23	24	25	26	27	28	29	30
INITIAL PREY (HARES)															
TOTAL PREY CAPTURED (HARES)															XX XX XX XX XX
PREY THAT SURVIVED (HARES)															XX XX XX XX XX
INITIAL PREDATORS (LYNX)															
SURVIVING PREDATORS (LYNX)															XX XX XX XX XX
PREDATOR OFFSPRING (LYNX)															XX XX XX XX XX

On the semi-log paper that is enclosed, draw two graphs: The Initial Prey and the Initial Predators (rows 1 and 4). Use two different colors of ink or pen or pencil or somehow differentiate between the two graphs.

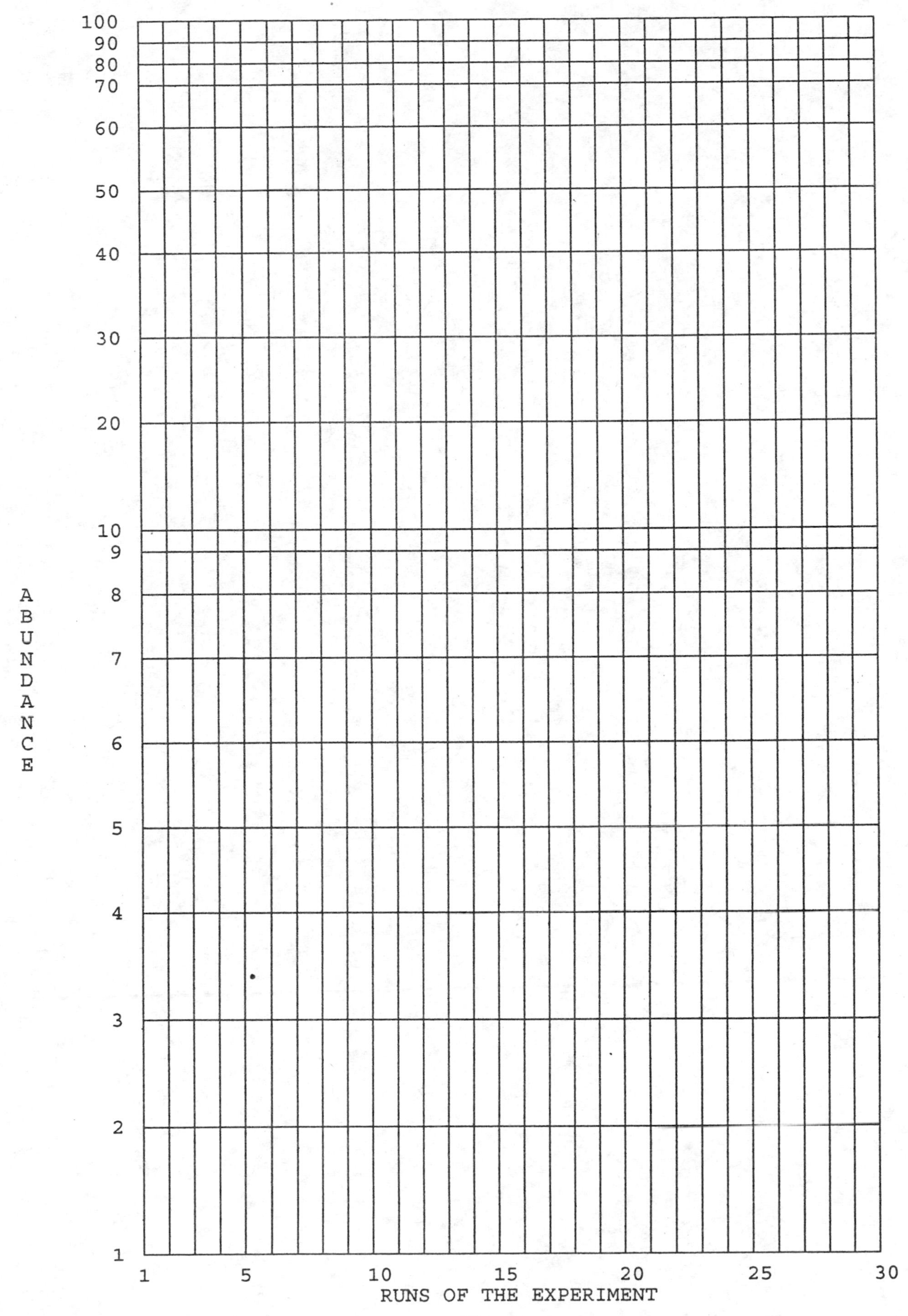
100
90
80
70
60
50
40
30
20
10
9
8
7
6
5
4
3
2
1
ABUNDANCE
1
5
10
15
20
25
30
RUNS OF THE EXPERIMENT

Name____________________

EXERCISE 9: SYMBIOSIS

Symbiosis is a close, intimate relationship between two different species of organisms. It is an interspecific interaction and infers a close association between the organisms where one organism lives on or in the other. This organism derives some benefit from its host. The host organism can be affected in a variety of ways. **Mutualism** occurs if the host also benefits from this obligatory relationship, that is both organisms must interact or there will be harm to both. **Commensalism** occurs when the host is not affected, neither benefits or is harmed from the relationship; if the host is harmed, the relationship is **parasitism**.

Although there are other types of symbiotic relationships this lab will be limited to exploring these three.

While it is impossible to discuss here, consider the evolution of a symbiont with its host, the **coevolution**!!

MUTUALISM

Lichens are symbiotic associations between fungi and algae. The algae live among the mass of thread-like hyphae of the fungi. The algae are photosynthetic and produce organic compounds used by the algae and the fungi. One of the functions of the mass of hyphae is to absorb these nutrients. The fungi are not photosynthetic but allow this association to grow on bare rock or other areas where organisms, previously could not adhere. Acids, the metabolic products of the fungi, break down the rock and produce **SOIL**!!

Obtain a slide of the cross-section of a **crustose** lichen. These lichens are scaly and grow closely against rocks and trees. Draw a sketch of this slide. Label the algal cells and the fungal hyphae.

Crustose lichens are the pioneer plants that grow on bare rock. This is an example of **primary ecological succession**. Primary succession occurs in areas that have not been previously changed by other living organisms, such as on bare rock; while lichens are usually the first to colonize an area, there are other mitigating factors that allow colonization by other organisms.

In the successional process, the environment is altered by the actions of the organisms and as a soil layer is developed, other organisms can colonize and thrive. Many times, crustose lichens are

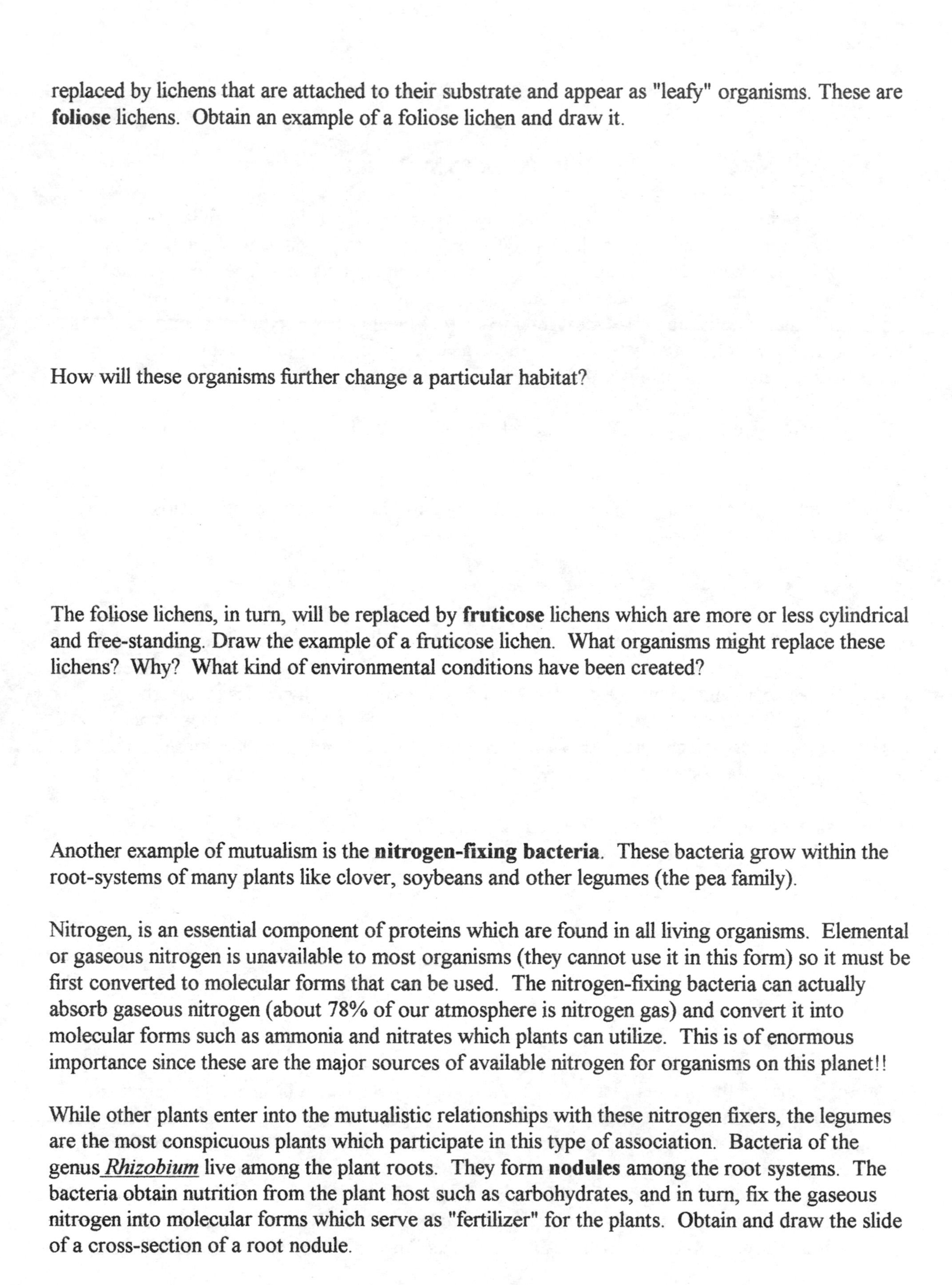

replaced by lichens that are attached to their substrate and appear as "leafy" organisms. These are **foliose** lichens. Obtain an example of a foliose lichen and draw it.

How will these organisms further change a particular habitat?

The foliose lichens, in turn, will be replaced by **fruticose** lichens which are more or less cylindrical and free-standing. Draw the example of a fruticose lichen. What organisms might replace these lichens? Why? What kind of environmental conditions have been created?

Another example of mutualism is the **nitrogen-fixing bacteria**. These bacteria grow within the root-systems of many plants like clover, soybeans and other legumes (the pea family).

Nitrogen, is an essential component of proteins which are found in all living organisms. Elemental or gaseous nitrogen is unavailable to most organisms (they cannot use it in this form) so it must be first converted to molecular forms that can be used. The nitrogen-fixing bacteria can actually absorb gaseous nitrogen (about 78% of our atmosphere is nitrogen gas) and convert it into molecular forms such as ammonia and nitrates which plants can utilize. This is of enormous importance since these are the major sources of available nitrogen for organisms on this planet!!

While other plants enter into the mutualistic relationships with these nitrogen fixers, the legumes are the most conspicuous plants which participate in this type of association. Bacteria of the genus *Rhizobium* live among the plant roots. They form **nodules** among the root systems. The bacteria obtain nutrition from the plant host such as carbohydrates, and in turn, fix the gaseous nitrogen into molecular forms which serve as "fertilizer" for the plants. Obtain and draw the slide of a cross-section of a root nodule.

Draw a sketch of the bacteria in the genus *Rhizobium* under the **demonstration scope**.

Draw a sketch of the preserved specimen of a legume root mass with the nodules.

What other organisms fix nitrogen? (Hint: we have looked at these aquatic organisms in a previous lab). What is the ecological importance of these creatures?

The last example of mutualism concerns the termite and its protozoan symbionts. Although termites eat wood, they cannot digest it!! The protozoans (members of genus *Trichonympha* and *Pyrsonympha*) digest the wood. The termites, in turn, serve as the habitat for the protozoans.

Examine the prepared slide of *Trichonympha* (termite flagellates) and draw a sketch.

Obtain the preserved specimens of the life cycle of the termite and draw the different stages. Note the damage that can be done to the wood.

COMMENSALISM

Commensalism can be illustrated in an variety of ways, few of which lend themselves to laboratory work. Epiphytic plants such as Spanish moss and tropical orchids grow on trees. They benefit by having a substrate on which to live but they do not benefit or harm the host organism. Observe the preserved specimen of Spanish moss. It really isn't a moss but a flowering plant. Where are its roots?

Our body has an infinite number of bacteria growing on it and in it. This is our normal bacterial flora and these organisms as a community, may be considered as commensals. They enjoy an almost optimum environment in the large intestine and other areas. Why might these harmless bacteria actually benefit their human hosts?

Observe two common bacteria of our body, *Escherichia coli* and *Streptococcus salivarius* under the **demonstration scopes** and draw sketches of them.

PARASITISM

Parasitism is a symbiotic relationship where one creature derives its nourishment at the expense of the other creature. Usually the parasite draws nourishment from the tissues of its larger host.

The host, in essence, is the habitat for the parasite. Parasites exhibit a great diversity. Viruses, bacteria and the more well-known protozoa and worms can be parasites. But there are parasitic plants, fungi and less well-known animals. There are microparasites (the tiny ones) and macroparasites as well as ectoparasites and endoparasites. List a few well-known ectoparasites.

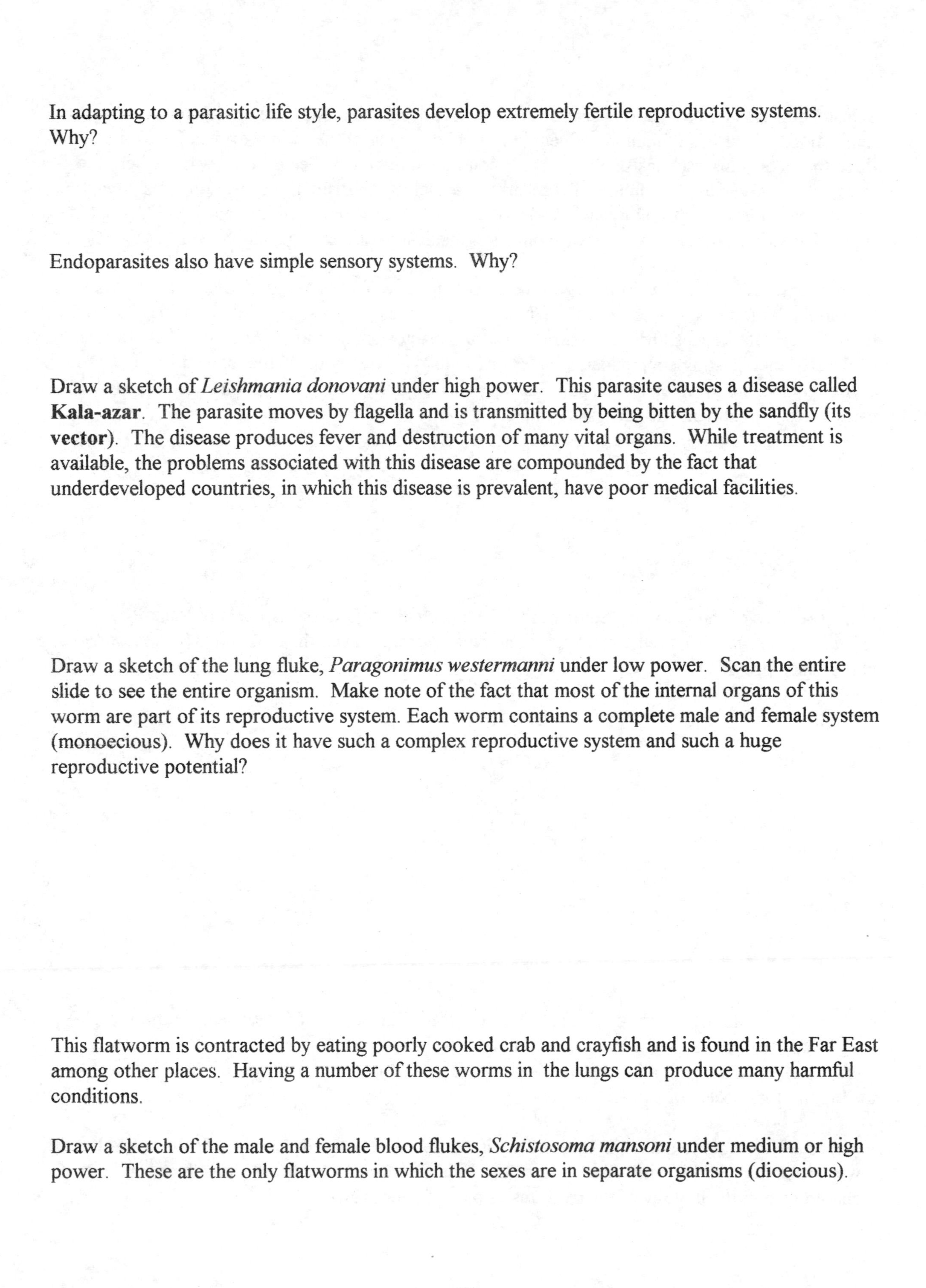

In adapting to a parasitic life style, parasites develop extremely fertile reproductive systems. Why?

Endoparasites also have simple sensory systems. Why?

Draw a sketch of *Leishmania donovani* under high power. This parasite causes a disease called **Kala-azar**. The parasite moves by flagella and is transmitted by being bitten by the sandfly (its **vector**). The disease produces fever and destruction of many vital organs. While treatment is available, the problems associated with this disease are compounded by the fact that underdeveloped countries, in which this disease is prevalent, have poor medical facilities.

Draw a sketch of the lung fluke, *Paragonimus westermanni* under low power. Scan the entire slide to see the entire organism. Make note of the fact that most of the internal organs of this worm are part of its reproductive system. Each worm contains a complete male and female system (monoecious). Why does it have such a complex reproductive system and such a huge reproductive potential?

This flatworm is contracted by eating poorly cooked crab and crayfish and is found in the Far East among other places. Having a number of these worms in the lungs can produce many harmful conditions.

Draw a sketch of the male and female blood flukes, *Schistosoma mansoni* under medium or high power. These are the only flatworms in which the sexes are in separate organisms (dioecious).

Schistosomiasis is a very serious disease because it is debilitating and chronic due to the ease of reinfection. Hundreds of millions of people throughout most of the third world suffer from the three forms of schistosomiasis. The worm is contracted merely by being in the water where the microscopic larvae are swimming. These larvae can actually burrow right through the skin and into the blood vessels. An individual does not have to eat or drink anything that is contaminated. Again, because of the poverty in these countries, medical treatment is primitive.

Draw the whipworm *Trichuris trichura* under medium power. It is found all over the world and lives in the large intestines of its host. While most parasites are host specific, that is they are only found in a single species, the whipworm parasitizes many animals, including man. The worm is contracted by ingesting contaminated water or food. While many times the infestation is asymptomatic (no symptoms), the parasite can produce diarrhea, bloody stools, loss of weight, nausea, etc. Treatment with drugs is available.

Draw sketches of the following ectoparasites: the Tse tse fly (*Glossina* spp.), the black fly (*Simulium damnosum*), the human louse (*Pediculus humanus*) and the crab louse (*Phthirus pubis*) in the space provided.

The first three can be vectors of disease. The Tse tse fly carries African sleeping sickness while the black fly transmits river blindness. Both flies also take blood meals from their prey. The human louse can transmit a number of diseases but usually just feeds, as does the crab louse, on the blood of its host.

Humans are the host to many parasites. Why is it that the developed countries of the world have fewer problems with parasitic diseases than the underdeveloped countries? Consider economics, technology, and the biology of the parasites as part of your answer.

Name______________________________

EXERCISE 10: SHELFORD'S LAW OF TOLERANCE #1

A cardinal principle in the field of ecology is the "Law of Tolerance." Synthesized by V.E. Shelford, the "law" states that the presence and success of an organism is controlled by a qualitative or quantitative deficiency or excess of some environmental factor. In other words, too much of an environmental factor as well as too little (Liebig's Law of the Minimum) can be limiting to an organism. So organisms have an ecological minimum and maximum, with a range in between which represents its limits of tolerance for that environmental factor $^{e.g.}$ temperature, light, water, oxygen. Exceeding these extremes will stress the organism; if the stress is great enough, the organism dies. Organisms try to operate at their optimum value for that environmental factor.

Organisms may have a wide range of tolerance for one factor and a narrow range for another. Organisms with wide ranges of tolerances for all factors are likely to be most widely distributed. Although the tolerance for one factor may not have any bearing on an organism's tolerance for other factors, if the organism is stressed, its tolerance limits to other factors could be affected.

For this exercise and the following one, we will see the effects of environmental stresses on organisms. We shall observe how these stresses (pollutants, if you will), affect the growth and reproduction of organisms. There are two variables in these experiments: **Some of the organisms will respond differently than others based upon that organism's tolerance limits and the other variable is the strength or duration of the environmental stress.**

The organisms that we use are bacteria. Bacteria grow rapidly, are fairly easy to work with and give accurate results. The instructor will demonstrate the **STERILE TECHNIQUES** that are necessary when working with bacteria. Although our bacteria are 99.0% harmless, it is necessary to use care and caution when working with them. Please do not bring food to the lab when these experiments are carried out. Furthermore, **IF YOU ENTER LAB LATE, AFTER THE INSTRUCTIONS ARE GIVEN, YOU WILL NOT BE ALLOWED TO PARTICIPATE; YOU WILL BE GIVEN A ZERO FOR THIS ASSIGNMENT. THERE ARE NO EXCEPTIONS!!!**

The lab schedule for today is the following:

Exercise 10: The effects of radiation--an experiment to observe the effects of radiation on bacterial growth.

and

Thermal pollution--Shelford's Law of Tolerance is illustrated by observing the growth of bacteria under different temperature regimes.

Exercise 11: Acid precipitation--an experiment which indirectly shows the effects of acid precipitation by varying the pH of the environment. There is also an exercise which will demonstrate the significance of pH on biological systems.

and

Toxic wastes--an experiment which illustrates the effects of different toxic substances on different bacterial growth.

Directions: Follow the instructor's directions **exactly!!!** If at any point in time you do not know what you are supposed to do, **STOP!!**, Ask the instructor. If there is any kind of accident, **STOP!!**, notify the instructor.

Some words about working with microorganisms:

STERILE TECHNIQUE must be adhered to when working with bacteria. A good rule of thumb is to sterilize everything before and after using it. This includes the inoculating loop, the lip of the culture tube and the tubes to be inoculated. Furthermore, the laboratory table top should be disinfected at the start of the lab and when the lab involving the bacteria is finished.

Microbiology is a relatively young science. It wasn't until the latter part of the 19th century that the "golden age" of the science began. Two discoveries helped usher in this "golden age." The first was the vehicle in which bacteria could be grown in - the **Petri** dish. The student will label the Petri dish in the following manner: Using the china marking pencil, label the underside of the Petri dish (the smaller of the two halves) with your initials, the name of the bacterium assigned and any other pertinent information. This is because Petri dishes are stored upside down-bottom half up).

The second innovation was the growth medium (food) on which bacteria would grow and reproduce. This was the discovery of nutrient **agar**, a derivative of seaweed, which at room temperature is a gel and when heated becomes a liquid. To this agar is added a variety of minerals and nutrients which serve as food for the bacteria since they are not photosynthetic.

The bacteria must be transferred to the Petri dish or nutrient broth tube from the culture tube in which they are growing and this necessitates **STERILE TECHNIQUE!!!**

In some of these exercises you will be working with a lab partner. In any event, have your lab partner help you. In essence, there will four hands to do these techniques instead of two. Follow these instructions:

1. Remove the cap from the culture tube but **do not** place it on the table - hold it between your palm and your pinkie.
2. Flame the lip of the culture tube for just a second or two in the Bunsen burner.

3. If a sterile swab is being used for inoculations, place it in the culture tube, squeeze off the excess before removing it from the tube. Now inoculate the Petri dish as per your instructor's directions.

4. If an inoculating loop is used, place it into the Bunsen burner until it becomes red hot and glowing. Remove it, let it cool for 10 seconds. Place it in the culture tube and remove a loopful of bacteria. You will know if you have a loopful because it works on the same is full. principle as a jar of bubbles that you played with when you were a kid, you can see if the loop is full.

ALWAYS FLAME THE LOOP AND THE LIP OF THE CULTURE TUBE AFTER AN INOCULATION TAKES PLACE!!!!!

5. If nutrient broth tubes are going to be inoculated make sure that the lips of those tubes are flamed before and after the inoculation.

Exercise 1. We are always being affected by radiation; from the sun, appliances, high tension wires, etc., etc. We still do not know what the effects of some of these types of radiation are. We know, for example, that sunlight and our continued exposure to it, can result in skin cancer.

In this lab, you will be assigned a specific bacterium and the time you must expose the bacterium to the radiation.

You will be inoculating a Petri dish using a sterile swab and you must swab the entire surface of dish-all of the agar. This can be achieved by rotating the Petri dish 90° four times as you swab. Make sure you do not break through the agar by swabbing too hard. Once you have swabbed the entire dish, remove the top of the Petri dish (the larger of the two parts, the part that fits over the other. Cover one half of the dish with an index card and expose the other half to the ultraviolet radiation within the light box for the exposure time that you have been assigned.

What has occurred is that after allowing bacteria to potentially grow over the entire dish, the half of the dish that is covered will, indeed, grow. The half that has been exposed may or may not grow depending on two things: **The time of exposure to the ultraviolet light and the tolerance limit of the bacterium to this type of radiation**. Record your results as well as the entire classes results, next week.

The two bacteria used in this experiment are *Staphylococcus aureus* and *Bacillus megaterium.* "Staph" is the very famous bacterium that causes Staph infections - boils, abscesses, toxic shock, etc., etc. *Bacillus* is a bacterium that is commonly found in the environment and there is no known pathology associated with it. Handle both bacteria with care!!

Bacterium: ____

Exposure Time: ____________________.

Results:		10 sec.	20	40	80	21/2min	5	10
	Staphylococcus aureus							
		1 min.	2	4	8	15	30	60
	Bacillus megaterium							

c) Conclusions:

Exercise 2. Electrical generating power plants, especially nuclear power plants use large amounts of water for cooling different parts of the plant. Many times this water, which enters the plant at a low temperature, is returned to the environment much warmer. Thermal pollution is the detrimental effects to organisms by raising (or lowering) the temperature. This lab will demonstrate the effects of temperature on bacterial growth.

Each pair of students will be assigned a bacterium and will inoculate four tubes of nutrient broth or Petri dishes with TSA media (agar). One tube or plate will be grown in a cold environment, one at room temperature, one at body temperature and one in a hot environment (5°, 25°, 37°, and 42° C., respectively). Note the effects of temperature on growth. The two bacteria used in this exercise are *Escherichia coli* and *Serratia marcescens*. *E. coli* is the very famous coliform bacterium that occurs in infinite numbers in the digestive tract of humans and other animals. *Serratia* is also an enteric bacterium that can be pathogenic but is also found out in the environment *E. coli* can be pathogenic in exceptional cases.

Bacterium ______________________________________

Results:

	5°	25°	37°	42°
Serratia marcescens				
Escherichia coli				

Conclusions:

Name________________________

EXERCISE 11: SHELFORD'S LAW OF TOLERANCE #2

Acid precipitation is a controversial and apparently increasing problem. As the pH drops (becomes more acidic) in a body of water or the soil, organisms will be stressed. Usually, the reproductive potential of an organism is affected first with gametes (sex cells) no longer being viable. The fertilized egg similarly is affected or as the embryo develops, birth defects occur. As the pH drops even further, adult organisms can die.

In the first part of this lab, each pair of students will inoculate three tubes of nutrient broth with the assigned bacterium. Each tube of broth is at a different pH. One tube of broth is acidic (pH = 5), one is neutral (pH = 7) and one is alkaline or basic (pH = 9). Note the effect of pH on bacterial growth. The two bacteria used in this exercise have been used previously, *Staphylococcus aureus* and *Escherichia coli.*

Bacterium ____

Results:

	5	7	9
Staphyloccous aureus			
Escherichia coli			

Conclusions:

After completing the next part of the lab, dealing with toxic wastes, finish the pH lab.

There are many toxic substances in our environment. Their effects, duration of exposure and the actual dangerous exposure levels are by and large, still unknown. In this experiment students will be assigned a bacterium and a toxic substance that the bacterium will be exposed to. Observe the effects of the toxic substances on bacterial growth.

Each individual will inoculate a Petri dish with the assigned bacterium and then place a sterile disk, impregnated with the toxic substance in the center of the dish. Next week there will be a zone of inhibition around the disk, where no bacteria grew. The size of this zone of inhibition will depend on two things: **the type of toxin used, and the tolerance limit of the bacterium to the toxin**. We will use the two bacteria from the very first exercise illustrating Shelford's Law of Tolerance, *Staphylococcus aureus* and *Bacillus megaterium*. The "toxins" we will use are: bleach, formaldehyde, listerine, phenol and the "mystery guest" (disinfectant).

Bacterium ____________________________________

Toxic substance _____________________

Results:	Bleach	Formaldehyde	Listerine	Phenol	"Mystery Guest"
Staphylococcus aureus					
Bacillus megaterium					

Conclusions:

The pH of a substance can be loosely defined as the acid-base balance or the measure of acidity or alkalinity of a substance. While there are technical definitions for pH, let's just leave it at that. It is important to note that all of the biochemical reactions that define an organism, that allow the organism to exist occur at a specific acid - base balance - a specific pH. If the pH shifts, and because the pH scale is exponential, shifts just slightly, these biochemical reactions **will not** occur. A slight shift in the pH represents a substantial change in the acid-base balance.

Humans and other organisms have built-in homeostatic mechanisms for maintaining a constant pH - buffers. An example of a buffering system that you are aware of concerns aspirin, an acidic substance, which upsets the stomachs of some users so it cannot be utilized for pain relief. Buffered aspirin (e.g. Bufferin) contains substances to counteract the acidity of aspirin so it can be utilized as a pain reliever.

The pH scale ranges from 0 to 14 with 7 being neutral, neither acid or basic. As the scale drops, substances become more acidic and the higher the pH, the more alkaline or basic a substance becomes. The scale is exponential with a pH of 6 being 10 times more acidic than 7, 5 a hundred times more acidic, 4, a thousand, etc. Similarly, a pH of 8 is 10 times more alkaline than a pH of 7, 9, a hundred times more alkaline, 10, a thousand, etc., etc. The scale is illustrated below.

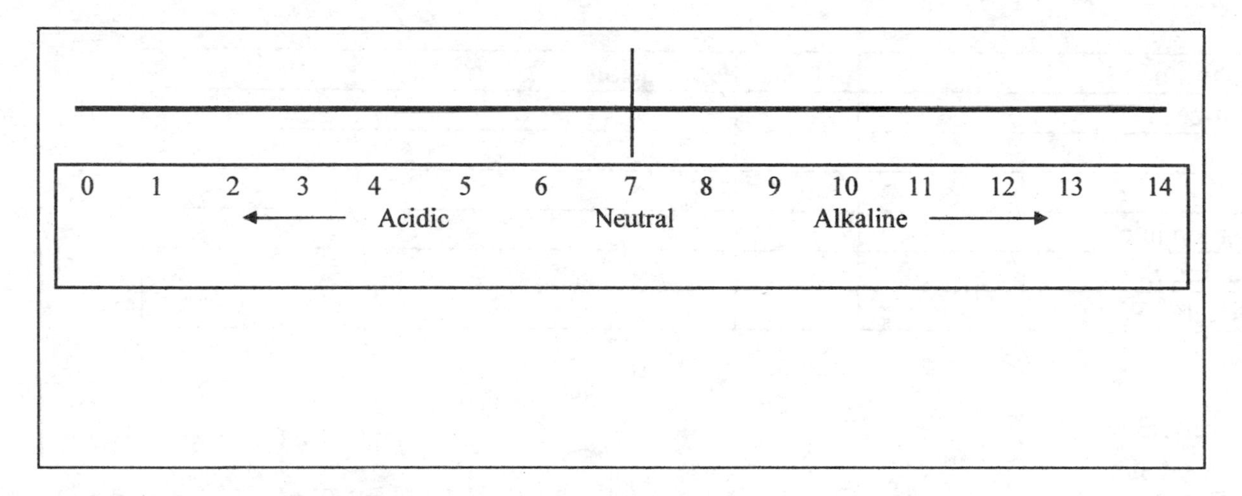

Exercise1: pH Determination using Litmus Paper.

Dip a strip of red litmus paper into each of the following solutions in Table 1 and record the results. Repeat the procedure using blue litmus paper. Look for a color change and use fresh paper for each determination. Acidic solutions turn blue litmus paper red and alkaline or basic solutions turn red litmus paper blue.

Table 1.

Solution	Red litmus Paper	Blue Litmus Paper	Conclusion
Coffee			
Cola			
Milk			
Lemon Juice			
Baking Soda			
Distilled Water			

Exercise 2: pH Determination using pH paper.

In a similar fashion as in exercise 1, dip pH paper into the different solutions. Match the color change on the paper to the **color chart on the pH paper dispenser**. Record the pH of each solution in Table 2.

Table 2.

Solution	pH	Conclusion
Coffee		
Cola		
Milk		
Lemon juice		
Baking Soda		
Distilled Water		

Exercise 3.

Repeat the same procedure as in exercise 2, using a hand-held pH meter. Simply immerse the pH meter in a small beaker containing the solution to be measured. Hold the "on" switch for approximately 10 seconds and read the pH of the solution on the meter. Make sure that the pH meter is immediately returned to the beaker containing distilled water after each use. Compare the accuracy of the meter with the readings from the pH paper. Record your data in Table 3.

Solution	pH	Conclusion
Coffee		
Cola		
Milk		
Lemon Juice		
Baking Soda		
Distilled Water		

Exercise 4. The effects of buffers on pH.

Make sure that the electrodes of the pH meters are always immersed in distilled water when not in use.

1. Immerse the pH meter electrodes in 100 ml of distilled water and determine the pH. Record your results: ___________.

2. Slowly add 0.05M HCl to the distilled water, drop by drop. Gently swirl the beaker after each drop. Record the number of drops of HCl needed to change the pH one unit ________.
How did the solution change (acidic or alkaline, lower or higher)?

3. Remove the electrodes from the solution and rinse with distilled water. Immerse the pH meter in a beaker containing 100 ml of pH 7 buffer solution. Add 0.05M HCl drop-wise to the buffer, swirling after each drop. How many drops were needed to change the pH of the solution one unit? _________. What conclusions can you draw from this exercise?

4. Repeat the above exercise (2 and 3) using 0.05M NaOH instead of 0.05M HCl. Record your data as above: # of drops of 0.05M NaOH to distilled water: ________.

# of drops of 0.05M NaOH to pH 7 buffer: _______.

How did the solutions change?

Conclusions:

Name________________________

EXERCISE 12: DELINEATING WETLANDS

In 1988, realizing the importance in protecting its remaining wetlands, the state of New Jersey passed a comprehensive freshwater wetlands protection act. It was modified in 1989 to protect a buffer or transition zone immediately adjacent to the wetlands. This transition zone, when required, varies in size from 50-150 feet depending on the quality of the wetlands it surrounds, and is determined by the New Jersey Department of Environmental Protection (NJDEP).

New Jersey wetlands (marshes, salt marshes, swamps, bogs and fens), as all wetlands in general, serve in recharging ground water, serve as sinks for pollutants, serve as flood control, are habitat for wildlife and serve as open space for recreation. Rather than mosquito-infested, dismal, worthless areas to be drained and filled, the importance of wetlands and their protection has finally been realized. The New Jersey Wetlands Act of 1970 regulates coastal or tidal wetlands.

Prior to 1988, the U.S. Army Corps of Engineers (USACE) had jurisdiction over development. While most states are regulated by the Corps of Engineers, New Jersey (NJDEP) has slowly assumed control over most of the responsibilities concerning these wetlands and as of December, 1993, has assumed complete jurisdiction in regulating wetlands. Today, New Jersey has the most stringent wetlands protection law in the entire country.

While environmentalists are generally very pleased with this law, not everyone is. Builders and developers are unhappy with the realization that they can no longer do anything they please with the property they own. They can no longer indiscriminately develop (destroy) these wetlands. Small property owners too, can no longer do as they please with these areas.

However, it isn't that all development ceases. It is now carefully regulated. With cooperation from all parties, the developer, the municipality where they property is located and the state, compromises can be reached. In most cases all parties benefit from this cooperation.

The state (NJDEP) will verify the work of a wetlands consultant and issue a Letter of Interpretation (LOI) which allows the developer to develop the property, staying out of the wetlands and minimizing impacts to the buffer or transition zone. In a number of instances, under certain circumstances, wetlands and transition zones may, in fact, be developed. Special permits are necessary for these activities and they are scrutinized quite carefully. In some cases, wetlands must be created if other wetlands are going to be destroyed. The rules and regulations get quite complex.

There are three criteria for determining wetlands: soil - is the soil classified as a hydric (wet) soil? Hydric soils have a number of characteristics due to the fact that they have been inundated or saturated for a certain period of time. Chemical processes, resulting from water logging change the color of the soil and these changes can be used to classify them. Vegetation is the second criterion. Plants are classified as upland (UPL), not present in wetlands, obligate (OBL), always appearing in wetlands and facultative (FAC), perhaps present in wetlands but also present in uplands. The third criterion is hydrology - water relations. Is there standing water present? Are there silt-stained leaves on the ground, the result of running water? Do the trees have water marks?, etc., etc. Hydrology is often the least understood criterion because in the spring, sometimes everything looks wet and in the late summer, everything looks dry.

For the state of New Jersey, the most important criterion is soil. Soil color will not change despite the time of sampling. The plants don't know where they are supposed to grow and sometimes wetland plants grow in uplands and upland plants grow in wetlands. Soils don't lie.

Wetlands generally include marshes, swamps, fens, bogs and comparable areas. The disturbance of these wetlands, including their filling is regulated by the NJDEP under the Freshwater Wetlands Protection Act. This Act mandates the use of the Federal Manual for Identifying and Delineating Jursidictional Wetlands which uses the aforementioned three parameter methodology to ascertain the presence and extent of wetlands.

METHODOLOGY

The NJDEP or an environmental consulting firm will undertake an investigation of a site to ascertain if the requirements for classifying an area as a wetlands has been met and to delineate these areas from the surrounding uplands. As well as the active field investigation, the following sources are also reviewed to identify potential wetlands areas:

a) The NJDEP Freshwater Wetlands Quarter-Quadrangle map of the area.
b) The USDA/SCS Soil Survey for the county in question (in this case, Passaic County, New Jersey).
c) The Federal Emergency Management Agency (FEMA) Flood Insurance Rate Maps for the Township in question (in this case the Township of West Milford, New Jersey). This source is no longer required by the state.
d) The United States Department of Interior Geological Survey Map for the quadrangle in question (in this case, Newfoundland, N.J. Quadrangle).
e) NJDEP GIS Resource Data CD-ROM, Series 1. Office of Information Resources Management.

The field investigations consist of numerous soil borings, observation of hydrologic characteristics and detailed vegetation inventories within the project site and on both the wetland and upland sides of the wetland delineation boundary. For the wetland, boundaries between the wetland and upland are then delineated on the ground by placing sequentially numbered colored-plastic flags or wired stakes at intervals so that the previous flags can be observed. These flags are placed at a maximum of 75 feet apart using the methodology described in the federal manual. They are paired samples illustrating conditions on the upland and wetland sides of the boundaries.

By using the aforementioned data sources, a general image of the study area is obtained and these resources are used as general guides.

Soil types are obtained with a variety of soil augers. Soil is obtained from a depth of approximately 0.00 - 20.0 inches and classified in accordance with the Munsell Color Chart which assigns a number-letter designation. Generally, soils with colors found in the first two columns on the chart are considered hydric..

Lists of the hydric soil series for New Jersey have been developed by county as well as for the entire state.

Vegetation is identified by visual inspection and using the classification system devised by the U.S. Army Corps of Engineers and the N.J. Department of Environmental Protection, evaluated as previously described.

The surface hydrology is determined by visual inspection as well as recorded data from soil surveys, historical data, floodplain delineations, and aerial photography and used to ascertain whether appropriate features exist to support a wetland habitat.

This informaion is included in data sheets filled out in the field during the investigation.

The New Jersey Department of Environmental Protection is now the primary regulatory agency of freshwater wetlands, overseeing draining, removal of vegetation, excavation and any alteration of the water table. The U.S. Army Corps of Engineers regulates only the placement of fill into wetlands. A consulting firm will offer an opinion as to whether a wetlands will be regulated as freshwater wetlands by the New Jersey Department of Environmental Protection.

Since 1 July 1989, the NJDEP requires a transition area around wetlands of "intermediate and extraordinary resource value." Wetlands of exceptional resource value are defined by the State as freshwater wetlands which discharge into trout production waters or which are documented habitats for state and/or federally listed endangered or threatened species. Wetlands of ordinary resource value are certain isolated wetlands (less than 5000 square feet in size and more than 50% surrounded by development), ditches, swales and detention facilities. Wetlands which do not fit either of the above classifications are defined as of intermediate resource value.

For this laboratory exercise, a wetlands delineation (determining the boundary between the uplands and wetlands and State Open Waters) will be simulated. Basically, work will be done from maps and diagrams. Slides of various wetlands and the "tools" used by a wetlands delineator will be presented before the lab actually begins.

During this exercise the student will learn to read maps and use a variety of reference material that is also utilized by the wetlands scientist.

On page 97 is Figure 1., a portion of the United States Geological Survey (USGS) map (Newfoundland, N.J. quadrangle) which contains a large wetland adjacent to and north of Cedar Pond, on Bearfort Mountain in the Township of West Milford.

A. Taped to the blackboard is the entire map. Consult this map and determine what the scale of the map is: How many feet is represented by one inch? 1" = ________ ft.

1. Wetlands are represented by this symbol \\||/ on topographic maps. Using a ruler, draw rectangles around the wetlands adjacent to Cedar Pond. Find the dimensions of the wetlands (now rectangles) in inches and use the map scale to convert the dimensions to feet.

2. Area of a rectangle is determined by multiplying the width x length. Determine the area of the wetlands in square feet ($ft.^2$). Divide each total by 43,560 $ft.^2$ to convert to acres. Divide the area in acres by 2.5 to convert to the metric unit of area which is the **hectare**.

3. Do the same for Cedar Pond.

4. Topographic maps have the elevation contours on them. Each contour line represent twenty feet in elevation. Determine the elevation (altitude) for the wetlands (between which contours?) and for Cedar Pond. Find a contour whose elevation is indicated on the map and count up or down (in 20' increments) until the contour around the wetlands and Cedar Pond is found.

B. Figure 2. (page 98) is a portion of the Federal Emergency Management Agency (FEMA) map of the area we are studying. It is important when doing a delineation to make sure that the study area is not on a flood plain or actually have a stream corridor as part of the development area. Taped to the blackboard is the legend which explains the zone classifications.

1. What zone are the wetlands adjacent to Cedar Pond in?

2. Using the descriptions on the map in the front of the room, describe what zones A, B, and C are. What is meant by the 100 year flood plain?; what is meant by the 500 year flood plain?

C. Figure 3. (page 99) is a part of the Soil Survey Map for this portion of Passaic County. The federal government has soil surveys for the entire United States!! Each table has a copy of the entire soil survey for Passaic County. The soils are abbreviated on the maps. Identify these soils by referring to the soil survey books.

1. Determine all the soils that are adjacent to Cedar Pond and list them. Then go to the "Hydric Soils List" for Passaic County and determine which of the soils are hydric (wetland soils). If the soil is on the list, it is hydric. If it isn't on the list, it is non-hydric (upland). After determining which soils are hydric, then list the soils that are adjacent to the hydric soils in the wetlands surrounding Cedar Pond.

2. **<u>Briefly</u>**, describe the hydric soils and the upland soils in the Cedar Pond area in a sentence or two. Use the descriptions in the Soil Survey of Passaic County books

3. In the field, soil color is determined by removing the soil from approximately 18" below the surface and comparing the color of the soil to the matching color on the Munsell Color Chart. The soil color is given a letter and number classification from the chart.

a. Using the county soil survey book, find the Munsell Color Chart classification for one of the wetland soils and one of the upland soils (look for the Munsell classification for the soil at a depth of 18 inches).

b. Soils are removed with a soil auger. There are many types of augers depending on the type of soil to be sampled. At the front table is a "Dutch" auger which works well in soils that aren't too "stony." Draw a sketch of the "Dutch" auger.

4. There is no way to determine the hydrology of an area without visiting it so we can't. But knowing that one of the areas is in the wetland, and the other is an upland, fill in the section on hydrology on the data sheets.

5. After seeing the slides of the vegetation, below is a list of plants, by category, that are found in wetlands and uplands. First, consult a copy of the National List of Plant Species that Occur in Wetlands for Region 1 (which is our region), that has been distributed. List the scientific name of the plant and its wetland classification. Use the two accompanying "Wetland Determination Sheets," one for the upland and one for the wetland vegetation (pps. 100 and 101). Fill in the section for soils using the soil survey books and the material you have already completed. Use your imagination to fill in the section on hydrology. Sign the bottom.

   We use Latin names (scientific names) for the plants because the common name for a plant may vary from one part of the country (or world) to another; the Latin name is constant. For example, in the eastern United States, a common marsh plant is called tussock sedge, but in other parts of the country, it is called uptight sedge. If the plant is referred to by its Genus species (Latin or scientific name), *Carex stricta*, then everyone knows what the plant is.

Wetland:
Trees - yellow birch, red maple, Atlantic white cedar, green ash, gray birch

Shrubs - highbush blueberry, spice bush, silky dogwood, swamp azalea, buttonbush

Herbs - Uptight or Tussock sedge, broadleaf cattail, skunk cabbage, Northern pitcher plant, Alaska gold thread

Upland:

Trees - northern red oak, American beech, white oak, sweet birch, sugar maple.

Shrubs - witch hazel, flowering dogwood, mountain laurel, maple-leaf viburnum, Allegheny blackberry

Herbs - Feather false Solomon's seal, partridge berry, stinging nettle, Virginia strawberry, wild Lily-of-the-Valley.

FINALLY, ON BOTH THE SOILS MAP AND TOPOGRAPHIC MAP IN YOUR LAB MANUAL, OUTLINE THE WETLAND AREAS ADJACENT TO CEDAR POND IN RED MARKER OR SOME OTHER COLOR TO MAKE THEM STAND OUT.

BIBLIOGRAPHY

Federal Interagency Committee for Wetland Delineation. 1987. Federal Manual for Identifying and Delineating Jurisdictional Wetlands. U.S. Army Corps of Engineers, U.S. Environmental Protection Agency, U.S. Fish and Wildlife Service, and U.S.D.A Soil Conservation Service, Washington, D.C. Cooperative technical publication. xxx pp.

Munsell Soil Color Chart. 1985. Macbeth Division of Kollmorgen Corp., Baltimore, Md.

Reed, P.B. 1988. National List of Plant Species That Occur in New Jersey Wetlands. USFWS, St. Petersburg, Fl.

Soil Conservation Service. 1982. National List of Scientific Plant Names. U.S. Dept. of Agriculture.

Soil Conservation Service. 1987. Hydric Soils of New Jersey. In cooperation with the National Technical Committee for Hydric Soils. U.S. Dept. of Agriculture.

Soil Survey of Passaic County, New Jersey. USDA, Soil Conservation Service.

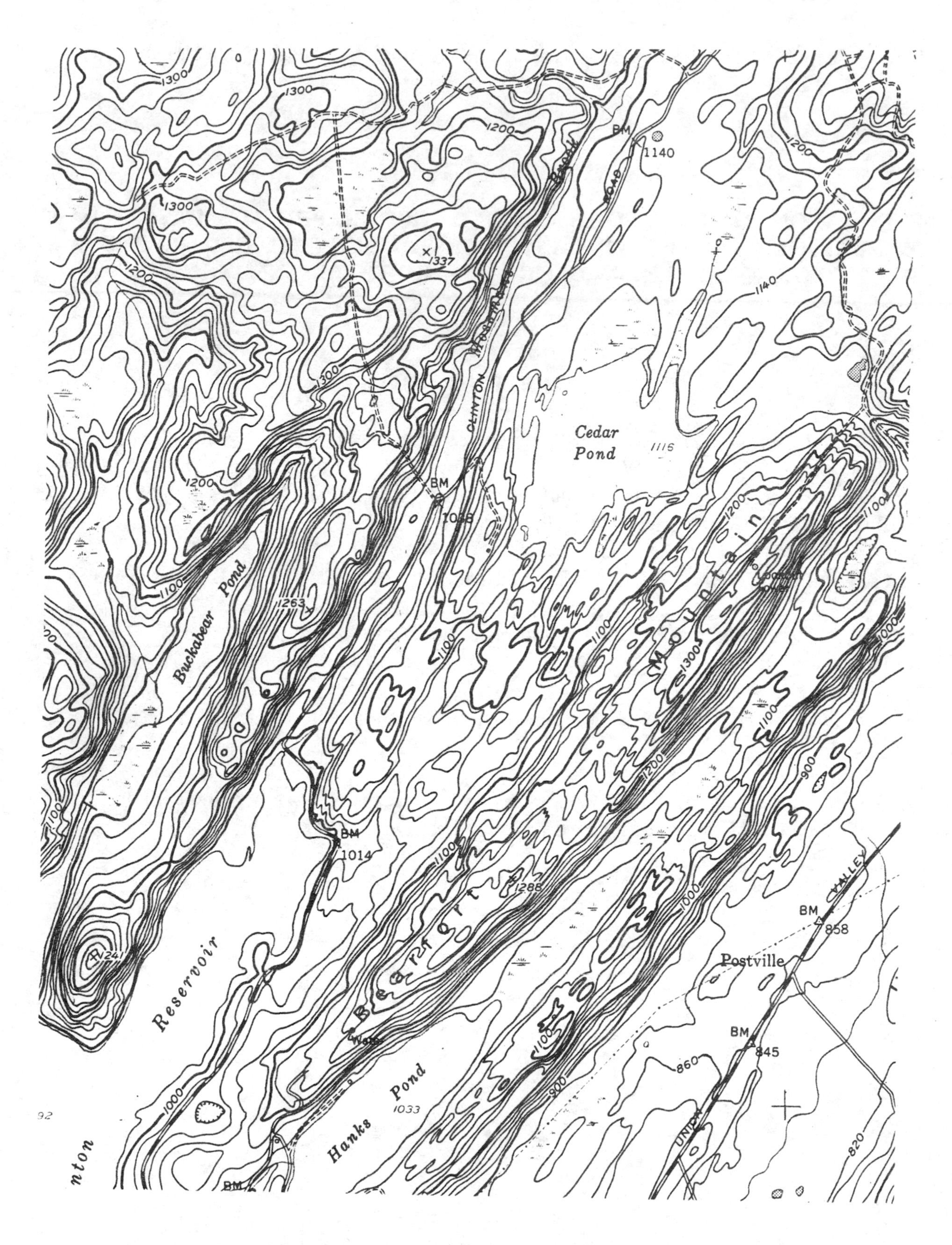

Fig. 1. Cedar Pond and adjacent wetlands, United States Geological Survey map, Newfoundland, N.J. quadrangle.

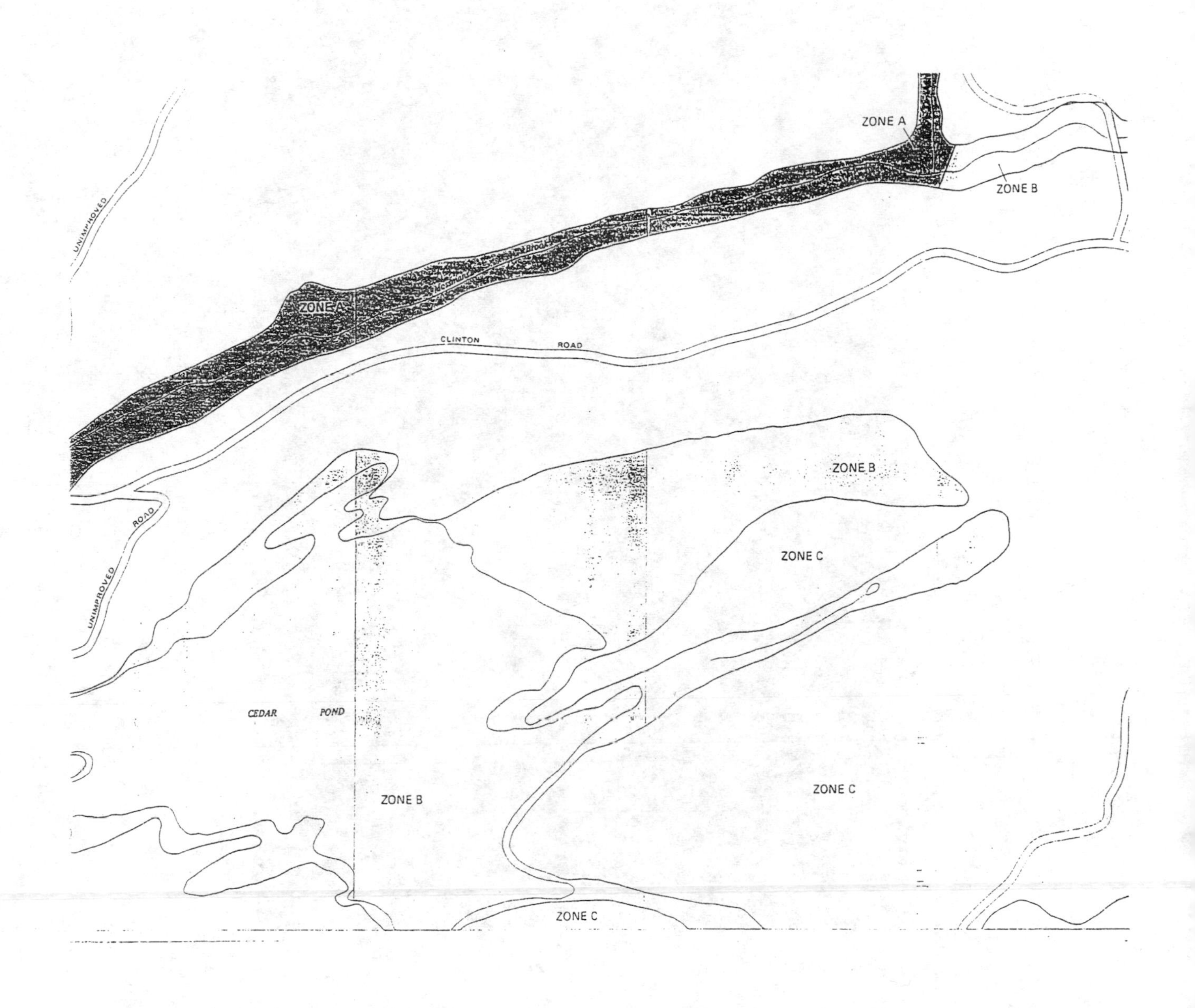

Fig. 2. A portion of the Federal Emergency Management Agency (FEMA) map of the Cedar Pond area.

Fig. 3. A portion of the Soil Survey map for this section of Passaic County, New Jersey, in the vicinity of Cedar Pond.

WETLAND DETERMINATION DATA SHEET

Applicant Name:________________________ Application Number:___________________ Project Name:_ ___________

State:__________ County:_______________ Legal Description_____________ Township:_______Range:__ Date:_____ Plot No.: _____
USGS Quad:________________ Quadrant:_______ Sample Location: N: ____ W:_____

Vegetation [list the three dominant species in each vegetation layer (5 if only 1 or 2 layers)]. Indicate species with observed morphological or known physiological adaptations with an asterisk.

SPECIES	Indicator Status		Indicator Status
TREES		**HERBS**	
1.		1.	
2.		2.	
3.		3.	
4.		4.	
5.		5.	
Saplings/Shrubs		**Woody Vines**	
1.		1.	
2.		2.	
3.		3.	
4.		4.	
5.		5.	

% of species that are OBL, FACW, and/or FAC:_____ Other indicators:________.
Hydrophytic vegetation: Yes__ No___. Basis:_____________________

Soil
Series and phase:_____________ On hydric soils list? Yes__ No .
Mottled: Yes____ No__. Mottle color___________________.Matrix color:_______________
Gleyed: Yes____ No____. Other indicators:________________________________.
Hydric soils: Yes____ No ; Basis:__________________________

Hydrology
Inundated: Yes_ No___. Depth of standing water:______________________.
Saturated Soils: Yes ___ No ___. Depth to saturated soil:_________________.
Other indicators: ____.
Wetland Hydrology: Yes ___ No __. Basis:________________
Atypical situation: Yes___ No___.
Normal Circumstances?: Yes_____ No____.
Wetland Determination: Wetland_______ Nonwetland ________.
Comments:

Determined by:________________________________

UPLAND DETERMINATION DATA SHEET

Applicant Name:_________________________ Application Number:____________________ Project Name:_ ___________

State:__________ County:______________ Legal Description______________ Township:_______Range:__ Date:_____ Plot No.: _____
USGS Quad:________________ Quadrant:_______ Sample Location: N: _____ W:_____

Vegetation [list the three dominant species in each vegetation layer (5 if only 1 or 2 layers)]. Indicate species with observed morphological or known physiological adaptations with an asterisk.

SPECIES	Indicator Status		Indicator Status
TREES		HERBS	
1.		1.	
2.		2.	
3.		3.	
4.		4.	
5.		5.	
Saplings/Shrubs		Woody Vines	
1.		1.	
2.		2.	
3.		3.	
4.		4.	
5.		5.	

% of species that are OBL, FACW, and/or FAC:_____ Other indicators:________.
Hydrophytic vegetation: Yes__ No___. Basis:______________________

Soil
Series and phase:_____________ On hydric soils list? Yes__ No .
Mottled: Yes____ No__. Mottle color__________________.Matrix color:_______________
Gleyed: Yes____ No____. Other indicators:____________________________________.
Hydric soils: Yes____ No ; Basis:___________________________

Hydrology
Inundated: Yes_ No___. Depth of standing water:______________________.
Saturated Soils: Yes ___ No ___. Depth to saturated soil:_________________.
Other indicators: ____.
Wetland Hydrology: Yes ___ No __. Basis:________________
Atypical situation: Yes___ No___.
Normal Circumstances?: Yes_____ No____.
Wetland Determination: Wetland_______ Nonwetland ________.
Comments:

Determined by:_________________________________

NAME________________________

EXERCISE 13: THE WOLF

Perhaps no other creature on this planet has been so misunderstood, persecuted and vilified by western civilization as the gray or timber wolf (*Canis lupus*). Once distributed throughout the Northern Hemisphere, the wolf has been extirpated from the vast majority of its former range. By the early part of this century, the wolf had all but disappeared from the continental United States, except for a small population in northern Minnesota. As we shall see, not only has this population survived, but with our new understanding and appreciation, has thrived and expanded its range into Michigan, Wisconsin and other adjoining states. It is believed that as of 1996, the wolf has reached its <u>carrying capacity</u> in Minnesota.

Why has the wolf been shot, trapped, and poisoned to the point of its disappearance in many places? The answer is very complex. It could be as complicated as western civilization's fear of the wilderness, of western religion's belief that humans have dominion over all other creatures and that the wolf represents evil or the devil, or something simpler. Wolves are **predators** and as such, are ferocious and inspire fear! As humans first came in contact with wolves, they were our partners in the hunt. This is probably how the wolf eventually was domesticated and became our **dog**. But eventually the wolf became a competitor with us for food and when humans began to raise livestock (sheep and cattle), wolves became a severe problem as they preyed on these creatures.

The wolf is the largest member of the dog family (Canidae), with males weighing up to about 100 pounds and females slightly smaller. They are about 5 feet long. Wolves are carnivores. They eat meat and their prey organisms are usually much larger than they are. They have an extraordinary sense of smell, which is at least one hundred times more sensitive than humans.

Wolves evolved about 60 million years ago from a primitive group of mammals called the creodonts and by 30 to 15 million years ago developed into creatures similar to the present day wolf. By one to two million years ago, the wolf as we know it today, had evolved.

While genetically identical to the dog, wolves have a slightly different skull structure, seldom have a curled tail like some dogs, breed only once a year unlike the twice per year cycle of the dog and have a precaudal gland that dogs lack. This gland is used for marking its territory.

Wolves generally live in packs - social structures with a hierarchy. There is an alpha male and female who are the dominants, and the other subordinate members. Members of the pack communicate with each other by physical contact, visual contact that is the most common form of communication, howling and other vocalizations, and a complex set of behavioral displays. Members of the pack actually form strong social bonds with each other. The pack has its own territory, which is delineated by scent marking and other means and is actively defended. Pack territory ranges from about 30 to 100 square miles (consider an area 10 miles by 10 miles - a relatively small area). Conflicts can arise between wolf packs with overlapping territories. When the wolf reintroduction program to Yellowstone Park finally was implemented, a political definition of a "pack" was necessary and this was defined as: at least two individuals that breed two years in succession. A pack usually ranges from two to over twenty individuals. When wolves become sexually mature at approximately 22 months, some members will leave their pack (which is, in many cases, is an extended family) and seek their own territories and mates (these wolves are known as "dispersers").

While wolves mate with a single partner for a relatively long period of time. Loss of a mate, leaving the pack, etc. mitigates this attribute. Both males and females share in caring for the pups. But in a pack, usually the alpha male and female are the only individuals that breed.

Wolves breed from late January through April depending on their location and many times have completed their den before the pups are born. The pups are born helpless, blind and deaf. They weigh about one pound. By three weeks of age they are outside near the den, playing and running. Wolves grow quickly so that by ten months to a year, they are essentially full-grown. They begin to learn their social skills and predatory skills soon after they are born.

The pups stay near the den until they are about eight to ten weeks old. While the adults are hunting, the pups will venture into the territory of the pack but will have specific "rendezvous sites" where they will stay until the adults return. These sites are used throughout the pup's first summer and into the early fall.

It has been shown that the chief prey of the wolf depends upon what large herbivores are present in their particular ecosystem. On Isle Royale in Lake Superior, the chief prey organism is moose, but whitetail deer, caribou, musk ox, elk, big horn sheep, bison, etc. can be the significant prey. Beaver and a number of smaller prey are also considered major food organisms.

Wolves often prey on the old, diseased and young because they are the easiest organisms to kill. The least amount of energy expended in capturing food has survival value. Why expend energy unnecessarily when survival depends on obtaining food in the easiest manner possible? Culling the old and the sick from a herd of herbivores can strengthen the overall population by removing those organisms and leaving the healthy and genetically robust to survive. However, there are data that suggest that increased mortality of young in a herbivore population can actually decrease the size of that population and jeopardized the survival of the population as a whole.

Watching a predator take its prey is not a pretty sight. A wolf pack exhibits many different strategies in killing the variety of prey organisms. It is interesting to note that animals that stand their ground to wolves (at least the large herbivores) have a better chance of surviving than organisms that run. Running seems to enhance the "rush" response of the pack toward the prey organism whereas when an organism stands its ground, the "rush" of the pack toward the "kill" is inhibited. Wolves have an uncanny ability to stop the chase if the chance of a capture and kill is unlikely.

Wolves are wild animals but they have an **innate fear of humans**. They **do not** make good pets because of this fear, nor do they make good watch animals. Wolf-dog hybrids are **not** good pets either because of their unpredictability. Consider a hybrid that has gotten the worst characteristics from each animal.

The issue of wolf reintroduction into parts of its former range is very controversial. The wolf is a predator, which tries to conserve energy by obtaining food, killing prey in the easiest manner. In Minnesota, wolf depredation on livestock is a serious issue. Although depredation of livestock is not a common occurrence, once a wolf has killed domestic animals, it must be destroyed because now it knows what an easy meal a cow or sheep is. The wolf must be killed to show ranchers and farmers that the state wildlife officials consider their concerns (as well as the wolf's concerns) important. Livestock losses are compensated for by a fund and the rancher must prove that the cause of death is by wolf depredation. This policy has worked well in Minnesota where ranchers and farmers tolerate wolf depredation knowing that they will be compensated for their losses. Similar programs are in place in the Yellowstone National Park reintroduction plan as well as the central Idaho reintroduction plan.

So the wolf is a complex predator, much maligned and misunderstood. Through education and clear discussions of all the issues, there is a place for wolves in the lower forty-eight states. All concerned parties must freely communicate with one another and reach viable solutions to the problems that wolf reintroduction poses.

At this point, in the lab, in order to gain more knowledge and a greater understanding of the wolf, slides and/or one or more of the fine videos available will be shown.

For more information concerning the wolf and wolf-human issues and to participate in the greater understanding of the wolf, you can join the International Wolf Center whose address is below:

International Wolf Center
1396 Highway 169
Ely, Minnesota 55731

Much of the information included in this lab comes from the work of Dr. L. David Mech, the world's foremost authority on wolves. The main reference is:

Mech, L.D. 1970. <u>The Wolf-The Ecology and Behavior of an Endangered Species.</u> University of Minnesota Press, Minneapolis.

The U.S. National Biological Service (NBS) tracks radio-collared wolves in and around Ely, Minnesota, from the air using a radio receiver with an antenna-a telemetry unit. Each collar on the wolves transmits a signal at a different frequency so that different wolves can be differentiated. As the airplane nears a collared wolf, the signal is picked up-a *beep*, which gets stronger as the plane gets closer to the wolf. Circling with the plane can virtually pinpoint where the wolf is located. Radio telemetry can also be done on the ground and wolves can also be located by visual sighting.

Accumulated telemetry data, that is readings taken approximately oncc pcr week can give valuable information about a wolf and the pack it represents. Each time a wolf is found using the telemetry unit, it is plotted on a map of the area. Each pair of students will use the *Center Section* map of Superior National Forest. These maps were obtained from the Forest Headquarters:

Superior National Forest
PO Box 338
Duluth, Minnesota 55801
(218) 720-5324

There may be large gaps in data from one year to the next due to budgetary constraints.

Each pair of students will be assigned a wolf (or wolves-they are numbered) and the telemetry data which have been collected. These data have been downloaded from the Internet and can be accessed from the Internet in two ways:

Through the World Wide Web at **http://www.wolf.org**
and via Gopher at **InforMNs.k12.mn.us**

Each pair of students will then plot all of the telemetry data on the laminated maps, with erasable markers. Mark the first year's data with small circles and the second with "X's". Use the black marker for the winter months of December, January and February, green marker for the spring months, March, April, May, red marker for the summer months of June, July and August, and blue marker for the fall months of September, October and November. If more data are available, use a different geometric shape. If your wolf tracking data stops abruptly, it may be due, for a variety of reasons, to the disappearance of the wolf. The wolf may leave the pack or its territory to seek a mate and form its own pack. Wolves also die from fighting with other wolves, starvation and of course, by humans, either purposefully or by accident. Check the data to see if your wolf reappears. Once all of the assigned telemetry data are plotted, draw a circular border around the plots.

Lets look more closely at the map of the C*enter Section* of Superior National Forest.

The map is divided by thicker lines (3" x 3") into different "Townships" which are numbered along the "y-axis" on the left-hand side of the map. It is also then divided into "Ranges" which is found along the "x-axis," on the top of the map. Within each "Township" are sections that are **numbered** and these sections can be divided into four quadrants: Northeast (NE), Southeast (SE), Northwest (NW) and Southwest (SW). The telemetry data that are available have "Township," "Range," and "Section" so that each pair of students can pinpoint the location of the wolf. Further more, cut-out templates, the size of a township (3" x 3") are supplied so that the exact locations on the map can be determined.

Below is a sample page of the wolf telemetry data. The U.S. Fish and Wildlife Service keeps track of approximately 30 wolves at a time. The data are updated regularly.

Wolf Telemetry Readings Sample

Wolf:	The number of the collared wolf M = male F = female
Time:	The time of the telemetry reading
TOF:	Type of Fix a = aerial v = visual g = ground
T.:	Township Number North (located on the left edge of the map)
R.:	Range West (located on the bottom edge of the map)
S.:	Section Number in Surveyed Township (1-36), guide located in legend of map)
DESCRIP:	Description of location
ACT:	Activity r = resting s = sleeping t = traveling f = feeding m = mortality
ASOC:	Associates, other collared wolves located in the same pack
TS:	Total number of wolves seen
CBM:	Could be more? Speculation about the existence of more wolves unseen

Wolf #	Date	TIME	TOF	T.	R.	S.	DESCRIP	ACT	ASOF	TS	CBM
75M	9/23/95	1034	a/v	61	12	9NW	0.5SE Blueberry L	r	423	5	Y
257F	9/23/95	1200	a	63	10	28SW	-	-	-	-	-

Here is some information about some of the radio-collared wolves that have been and are being followed near Ely, Minnesota. This information was downloaded from the International Wolf Center's Home page on the Internet. The data were collected by the National Biological Survey.

Wolf Number	Sex	Approximate Age	Pack	Pack Size	Weight
75	M	8.5 yrs.	Kawishiwi Lab	2	82 lbs.
141	M	7.0	Big Lake	4	92
253	M	4-7	Pike Lake	2-10	80
257	F	3-6	Birch Lake	3-6	54
381	M	5	-	?	91
451	F	4-5	Kawishiwi Lab	2	62
481	F	6-8	Pike Lake	2-10	68
505	F	3-4	-	?	79
517	F	3-4	Ima Lake	3	68
527	F	3-4	Pike Lake	2-10	56
531	M	3-4	Nina Moose	5	86
553	M	3-5	Moose Lake	2-3	82
555	F	2	-	?	53
567	M	Pup	Birch Lake	8-10	38
569	F	Pup	Birch Lake	8-10	31

Once all of the data are plotted and using the information that you have been given during this laboratory, answer the following questions:

1) Obviously, wolves do not stay in one place. What factors would cause them to move?

2) What does the entire area of movement of one wolf represent?

3) Discuss why wolves move during the different seasons.

4) Would there be any difference in movement if a wolf was male or female?

5) Wolves have been successfully reintroduced into Yellowstone National Park and in central Idaho. What issues are involved in the reintroduction of wolves to areas that they were once native to? Consider the issues if they were reintroduced into the New Jersey Pinelands, and the Adirondacks of New York State.

Name________________________

EXERCISE 14: AN INTRODUCTION TO THE COMPUTER AND SIMLIFE©™ [1]

Some students are quite knowledgeable in using a computer. Some are completely inexperienced and even intimidated by it. This lab is designed to introduce the computer to students and allow them to start utilizing this important tool. As well as using the computer for a variety of short exercises, we introduce an exercise that will continue next week and will essentially be a semester project.

On a personal note, the author is old enough to be able to say that he didn't have hand-held calculators during his college days, let alone computers. We were allowed to use a slide-rule for various arithmetic functions which gave us **approximate** answers. These **approximate** answers were considered sufficient for our work!!! To try and explain exactly what a slide rule is, is beyond the scope of this exercise. Suffice it to say that these days, they can only be found in antique shops and in the Smithsonian Institute.

Your author spent two years (count 'em) simply doing rewrites of his doctoral dissertation on a typewriter (a typewriter is also an antique). With the use of a word processing program loaded into a computer, your author could have accomplished the rewrites in a few months!!

Students are continually being asked to submit papers that are "typewritten" rather than in long hand. These documents can be saved and stored on a floppy disk and printed out at any time in the future.

In the early days of personal computing, a user may have been required to know how to program the computer. Today that is unnecessary. All the important programs (e.g. word processing) have been written. The student must merely know how to access the program in the computer and learn how to use the program. Also of utmost importance is the fact that an individual can play many, many, games on the computer which will waste an incredible amount of time and prevent the necessary work from being done promptly.

What we are going to do for this exercise is get the student up and running on the computer, eliminate any fears about computer operation, give some general instructions about some programs and let the student play with the computer for a while. We will then introduce next week's lab which will also become a semester project.

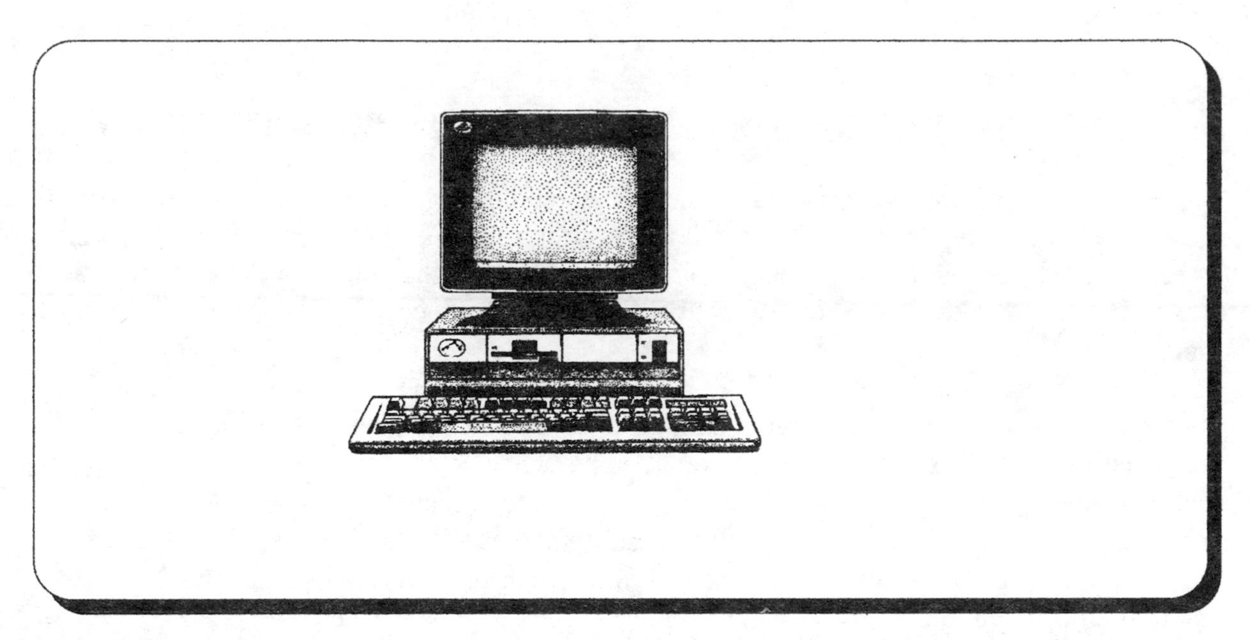

The above figure represents a "typical" computer. It consists of a keyboard, a monitor and the computer itself. Note the slot where the floppy disk is inserted. Floppy disks (see the diagrams below) can hold programs and data (e.g. reports). Most computers, these days, have a hard drive which is inside the computer chassis. Hard drives can hold a tremendous amount of information including all of the programs and all of your data. Specific programs and information can be loaded into the computer from the hard drive at your commands.

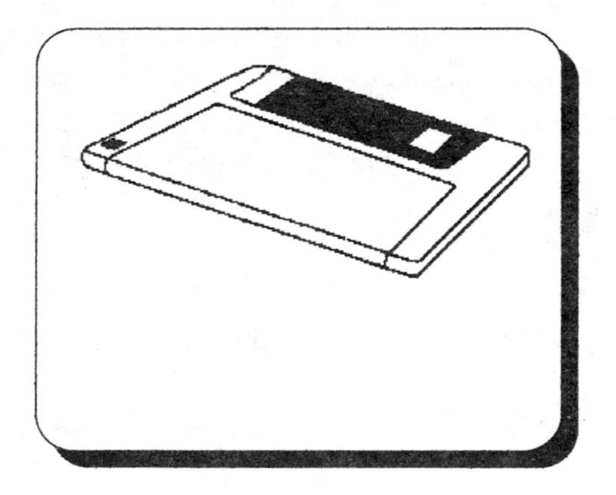

3 1/2" Floppy Disk

5 1/4" Floppy Disk

First, turn the computer on. The assumption is being made that the student is either working right in the lab or in the computer center or some similar venue.

Usually, all the programs are loaded into a hard drive in your computer. When you turn the machine on, a "C:\>" prompt will appear on the screen and you will have to type in the command of the specific program to run it. If a "menu" appears after the machine is turned on, all you have to do is press the number (or letter) on the keyboard that corresponds to the program you wish to run..

Although you may have no knowledge of how to use a word processor, it is simple to learn. The stigma attached to using a computer is highly overrated and learning the basics of word processing can be accomplished in a short period of time. Forget about the manual, forget about hours of training. We are going to use it right now, write a document and even print it out. So there!! It really doesn't matter which word processing program is loaded into the computer, you can do it. First some background:

Generally word processors fall into one of two categories, those that operate from DOS and those that operate from Windows®™[2]. DOS word processors are fast becoming obsolete because Windows applications allow much more flexibility and are mouse driven. DOS stands for **DISK OPERATING SYSTEM**. If you are using a Macintosh this lab is useless, go home turn on the T.V., take a break. DOS tells your computer how to operate, run programs, store data, etc., etc.

Generally, the **"FUNCTION"** keys (located usually at the top of the keyboard, F1, F2, F3, etc.) will bring down "menus" for use with a DOS-based word processor. Try them. Note what commands are located after each function key is pressed. Look for a command such as "create document," or "new." This will allow you to create a new document and start typing. If the command isn't as described as above, do a little experimentation to see how you can start typing a document. You really can't hurt the machine. Pressing the escape key generally brings you back to the start of the program. If you have found "create a new document" or something to that effect, start typing your notes from the most recent lecture in Environmental Biology. After you finish typing your lecture notes, look for the function key that will print your document and follow the instructions on your screen. Really, this is a no brainer. Before you print, make sure the printer (see below) has paper in it.

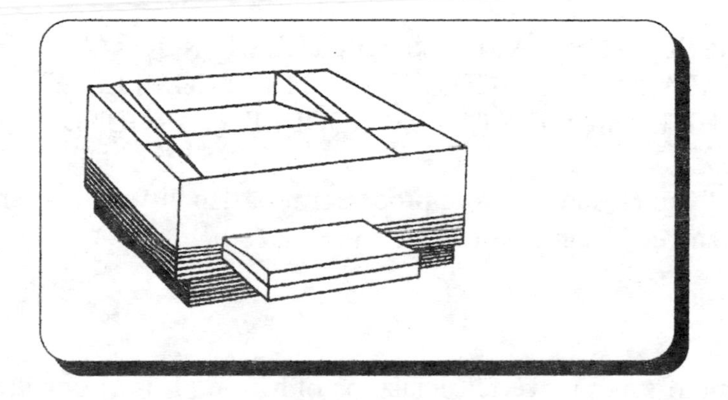

[2] Windows is a registered trademark of Microsoft Corporation.

It is also possible that your DOS word processor can be manipulated using a mouse (see below). A mouse will allow you to access different commands of the word processing program simply by pointing the cursor (it is an arrow) at the specific command and then clicking the left mouse button. Since all word processors are different. You will just have to experiment.

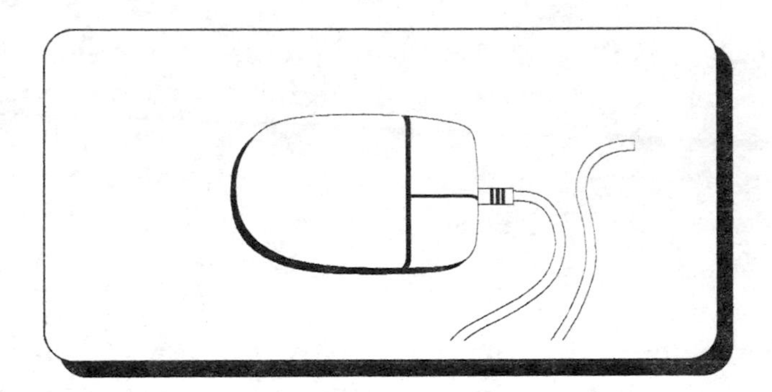

This lab manual was prepared using a word processor that works within the operating system called Windows. For years I used a very powerful, but simple word processor that was a DOS program. It served me very well, did everything I asked of it **but**....... When I wanted to get fancy. When I wanted to change the type (called fonts), when I wanted to place pictures (as above), I had to move to a Windows-based program because a DOS program couldn't get that sophisticated.

The reality is that today, most IBM compatible computers use Windows as the operating system and when the computer is turned on, Windows loads and runs automatically or there will be a menu that has Windows as one of its choices (press that selection); or at worst, the "C" prompt appears, and "Win" must be typed in. Windows is a system that uses a graphical interface---little pictures (**ICONS**). Click on the icon that is desired, in this case the word processor, and it loads. Windows programs are all icon driven and you must point to the command that you desire with the mouse and click the left mouse button. Experiment with the mouse and the commands. Note that if you click on "File," you will be able to "open" an existing file or with "new," create a new file. Create a new file. Start typing, type your last lecture. Find the icon for printing and print out your work.

WHILE YOU MUST CONSULT THE MANUALS FOR THESE PROGRAMS TO FULLY UNDERSTAND THEIR CAPABILITIES AND UTILIZE THEM COMPLETELY, YOU CAN REALLY FIGURE OUT THE BASICS JUST FROM NOODLING AROUND.

Now don't tell anyone, but if you close your word processor program in Windows and click on "Games," you can waste an incredible amount of time.

SIMLIFE

Simlife is described by its authors as a powerful simulation of nature. It is a computer model in which many concepts of biology can be illustrated. It can be used to demonstrate principles of genetics, evolution and most importantly to us, ecology and environmental biology. With Simlife we can design ecosystems and manipulate the chemical, physical and biological components. We can design organisms to be successful or unsuccessful in the ecosystems we build. Mutations can be introduced to illustrate change in organisms in response to environmental change. The options are almost limitless and by learning how to use Simlife and then designing simulations of your own, you will truly understand the important ecological principles that govern organisms and ecosystems.

Since "time" is an important concept in this course, and we constantly observe through time, Simlife can show the changes in our experiments over time.

Using Simlife, many of the important ecological concepts that have been, and will be discussed can be demonstrated such as: components of ecosystems, competition including predator-prey relationships, ecological succession, etc., etc.

In future exercises, the student will run an experiment in which the physical nature of the environment is changed and observe the effects on organisms within that environment. The student will run some of the built-in simulations (scenarios) in Simlife and collect data and interpret these data. Finally, the student will be able to design an original experiment - design an ecosystem with all of its components and then let it evolve over time to see if it is a successful environment where organisms thrive and maintain themselves.

For this exercise we will familiarize ourselves with Simlife by running the built-in tutorial. This fairly long exercise will expose the student to many of Simlife's capabilities.

Turn on the computer and at the "C" prompt type "Win," which will start the Windows program. When the "Program Manager" screen appears double click (with the mouse) the "Maxis" icon and then the Simlife icon. The program will begin. When the introductory screen appears, click once with the mouse to continue to the next screen. This screen will have a large dialogue box (one with many words). Click on the "Tutorial" box.

Read this box as an introduction. Then click on the "play scenario" box at the lower right.

When the "play scenario" box is clicked, the physical world of your scenario is being created. Watch it. A dialogue box appears.

Now, with the mouse, click "click here to continue." As you wait for the next message note that "time goes by" in the upper left hand corner of the screen. After reading the next message, again, click "click here to continue." Time goes by.

HERE IS THE TRICK TO GET THROUGH THIS TUTORIAL: <u>AFTER</u> READING THE INSTRUCTIONS, CLICK "CLICK HERE TO CONTINUE," **<u>THEN</u>** DO THE TASK THAT THE INSTRUCTIONS ASK YOU TO DO!!!! THE TUTORIAL WILL NOT CONTINUE UNTIL YOU DO THE TASK THAT IS ASKED CORRECTLY!!!

This tutorial takes between 60 and 90 minutes to complete and many times you will have to wait a long time (a minute or two, but it seems like a long time) before the next set of instructions comes on the screen. Your instructor will show you some short cuts once the tutorial is up and running.

The tutorial is in two parts. Each part is to be completed and the entire class will proceed with this tutorial together (more or less). Be **PATIENT**!!!! If you get lost inform your instructor immediately so that you don't fall too far behind. In the worst case scenario, you can start the tutorial over from the beginning.

Name_______________________________

EXERCISE 15: SIMLIFE #2

We have run the tutorial for Simlife and will run some of the scenarios. We have learned how powerful this simulation program is, as well as its limitations. We now will construct a world, set the physical parameters and run an experiment to show how variations in the physical characteristics of an ecosystem effect the organisms. Later, we will add more components to the worlds we create and see if we can sustain them. Future exercises will entail creating specific environments and creating new creatures that will be successful in those environments. By this time you should be somewhat comfortable using Simlife.

Let's create a new ecosystem by establishing the abiotic components. We can control the following parameters: temperature, moisture and soil depth. Together, these three components really allow us to establish the **climate** of this system.

Turn on the computer, start Windows, click the Simlife icon and making sure you are in the "Experimental Mode," click on "Play a Scenario." When this occurs, a "World Design" screen appears. You will now establish the basic characteristics of the ecosystem.

Using the mouse, set "Regional Weather Variation" to low (move the slider all the way down). Do not modify world average temperature or moisture (for now) and move the sliders so there are no mountains or rivers and lakes (set to "none").

Change the world size to "tiny." This makes your world 64 tiles across by 32 tiles down (the tiles are not well defined on your screen but you can make them out).

Change the new world name to your name or your lab partner's. Move the mouse cursor to the "New World Name" box and use the delete key to remove the present name and type in the new name. Click "Make It So." At this point we are going to save the world to a floppy disk. Insert the floppy into the disk drive. With the mouse click, in the upper left, "File " Click "Save As" and click "a" for the "a" drive. Under "File Name" you will see "a:*.life." Use the cursor, delete the asterisk and get the box to read: a:yourname.lif Click "O.K." and your world will be saved to the floppy disk. You should see the light by the disk drive go on. You will then be returned to the simulation. We will be saving our game very frequently but from here on, all you will have to do is click "save" (after clicking "File") and the game will be saved to the floppy disk.

Click on the minus sign in the upper left hand corner of the "Map of the World" window and then click "Close." Click the minus sign in the upper left of the window with your name and then click "maximize." The window fills the screen. The screen cannot show the entire world but you can "scroll" around using the left and right arrows on the bottom of the screen and the up and down arrows on the right hand side of the screen.

Go to the dashboard for Simlife (on top where all the icons are) and click the icon that looks like boulders or rocks (on the right side). This is the barriers tool and the **cursor will turn into a**

rock. We are going to divide our world into thirds. Since the world is 64 tiles across, starting at the top left, click the left mouse button 21 times and 21 boulders will be placed along the top. Then place boulders all the way down through the world. Remember that the screen doesn't show the entire world so when you get to the bottom, the screen will continue to scroll down or you may have to use the arrows previously mentioned.. If you note the "square" between the up arrow and down arrow on the right, when that square meets the bottom arrow, the end of the world (the end of the world!!) is reached. In Simlife the world is flat!! Use the cursor and go back to the top of the world. From tile 21 count 21 more tiles to the right, placing the boulders as you go. When tile 42 is reached, place boulders down the world once more. The world has been divided into thirds. Complete this part of the exercise by placing boulders around the entire margin of the world. Save the world to disk simply by clicking the file menu at the upper left and then clicking "Save."

We are going to manipulate our world now and keep it as simple as possible. Above the dashboard is the "Windows" menu. Click it and then the Climate Lab submenu. With the mouse cursor, move each slider all the way to the left. We are eliminating any variations in day length, rainfall and temperature. Note that the graphs on the right become straight lines. Close the climate lab (the box in the upper left of the window), click "Close." Save your world.

Click the Windows menu above the dashboard and click, "Map." The map of the world will appear and you will see it has been divided into thirds. Note that when you bring up the map window, the dashboard changes slightly. Look at the lower left of the dashboard, there are 5 icons, 4 of which are "off," one (altitude) is "on." To the immediate right of the icons is a scale for that particular parameter with colors indicating "low" to "high."

The first abiotic characteristic we are going to manipulate in your world is soil depth. We want one third of the world to have **deep soil**, the middle third, **moderate soil depth** and the last third to have **shallow soil.** Click on the soil depth icon (a root in the soil, the icon to the extreme left). Note the scale to the right of it, black indicates deep soil, olive to blue, medium and gray indicates shallow soil. We want the left third of the world black, the middle, olive to blue and the right third gray to blue. Now, click the Windows menu and then the Edit submenu. Hold the **shift key** down and type **WEIS**. The map window disappears and your world (your name is at the top) is present. Soil depth is indicated by the little squares. By holding down the left mouse button and scrolling around each third of the world, soil depth can be altered. The altitude icon is now used for soil depth (that's what typing WEIS does). So click the up arrow by the altitude icon to increase soil depth and the down arrow to decrease soil depth. Scroll around each third of the world and create the different soil depths that have been indicated. Remember that the screen is too small to hold the entire world so use the up and down arrows to the right of the screen and the left and right arrows on the bottom of the screen to scroll around your entire world. SAVE YOUR FILE!!!!

Return to the top left of the world using the scrolling arrows. Now each of the three **columns** must be subdivided into three **rows each.** Beginning at the upper left, count 11 tiles **down** and then using the barriers icon (the boulders in the upper right of the dashboard), place the barriers all the way **across** the world. Return to the left side of the world, count **down** 11 more tiles and place boulders across. Each column has been divided into three **rows**. Save your work.

We are going to make the top row of each column an area with high temperature (so, the top third of the deep soil will be hot, the top third of the moderate soil will be hot and so will the top of the shallow soil environment). Click on the temperature icon at the bottom of the dashboard (the thermometer). The cursor also turns into a thermometer. The up arrow by the icon will increase the temperature and the down arrow will decrease the temperature. The little squares that indicate temperature also appear on the world. Note the scale to the right of the temperature icon. Maroon indicates the warmest temperatures, yellow or orange is moderate and white is the coldest. Click the left mouse button so that the **top row of each column** becomes maroon. Holding down the shift key while clicking will speed up this process. Now move to the **middle row for each column** and using the up arrow for the temperature icon (or the down arrow) create moderate temperatures for the middle row for each column. Finally, move to the **bottom row for the three columns** and change all the squares to white, indicating a cold environment. You will probably have to click the down arrow by the temperature icon. Save your work.

Return to the upper left corner of the world. Now we will establish the hydrology (water relations) of your world. We are going to divide each column (deep, moderate, shallow soil into thirds (for a grand total of nine columns). In Simlife, the "water drop" on the dashboard controls the ability to make an area, wet or dry or something in between. In Simlife, the water relations mean rainfall and humidity. We are going to carry it a step further by allowing this characteristic to represent soil moisture, as well. We are going to divide each **column** into thirds using the same technique as previously used-placing boulders vertically down through your world. Your world is 64 tiles across and you previously divided the world into three columns, counting 21 tiles and placing the boulders down. If you start at the upper left, count 7 tiles (you can count the boulders now) and using the barriers icon (click it on), place boulders down the world. Return to the top of the world, count 7 more boulders to the right and place boulders down the world. Your original column has now been divided into thirds. Repeat this process for the **original** middle and right columns-divide them into thirds. The world now has 27 "boxes." We now want to make the first third of each column (don't get confused now, think!!, the first third of each column), a very wet - hydric environment. Click on the moisture icon - the "drop". Turn the boxes in the **first third of each column** as dark green as possible. Now for the **middle third of each column** we want the boxes shades of green to yellow to indicate a mesic (moderate water relations) environment. Remember that the up and down arrows by the drop icon will help. Do not be concerned if all of the boxes within the columns aren't the same color. As long as the great majority of them have the desired characteristics, that is sufficient. Now the **last third of each column** is to reflect a xeric environment. Click the down arrow by the "drop" icon and get all the boxes (or the vast majority of them blue). SAVE YOUR WORK!!!!!

There are now 27 boxes that comprise the world. All twenty-seven vary in terms of their physical environment. Let's review: You have created a world in which you have varied three abiotic parameters: soil depth, temperature and hydrology (which is really moisture and humidity in Simlife).

This is what the world looks like:

Deep Soil Hot Hydric	Deep Soil Hot Mesic	Deep Soil Hot Xeric	Moderate Soil Hot Hydric	Moderate Soil Hot Mesic	Moderate Soil Hot Xeric	Shallow Soil Hot Hydric	Shallow Soil Hot Mesic	Shallow Soil Hot Xeric
Deep Soil Mod.Temp. Hydric	Deep Soil Mod.Temp. Mesic	Deep Soil Mod.Temp. Xeric	Moderate Soil Mod.Temp. Hydric	Moderate Soil Mod.Temp. Mesic	Moderate Soil Mod.Temp. Xeric	Shallow Soil Mod.Temp. Hydric	Shallow Soil Mod.Temp. Mesic	Shallow Soil Mod.Temp. Xeric
Deep Soil Cold Hydric	Deep Soil Cold Mesic	Deep Soil Cold Xeric	Moderate Soil Cold Hydric	Moderate Soil Cold Mesic	Moderate Soil Cold Xeric	Shallow Soil Cold Hydric	Shallow Soil Cold Mesic	Shallow Soil Cold Xeric

Go to the Window menu and click "Map," the world can be seen divided into the twenty-seven boxes. Click the soil icon and the differences in soil can be seen. Then click temperature, observe the differences and then click moisture and observe. Save your work again, just to make sure.

Before we start manipulating the biotic components of this experiment, do the following: Go to the Simulation menu, click it and click the "Technical" submenu. Click "Change Physics." On this screen, to the right, is "Simulation Stuff" and within it, "Mutation Rate." Click it so the rate is <u>zero</u>! Click "make it so" and save your world.

We are going to observe the effects of the physical environment, as we have created it, on seed germination. We are going to select 3 species of plants, place 10 seeds of each plant in each of the 27 boxes and let them grow (or not). We will periodically stop the experiment and collect and record seed germination data. After the experiment is over you will answer questions based upon the data and draw some conclusions. If you stop and think about the physical parameters we are working with and the plants we've chosen, using common sense can lead you to the outcome of this experiment. Let's begin.

The three plants we are going to use are: Aloe, Balsam Fir and Sphagnum Moss. By selecting the these plants, we have selected a plant that is adapted to the desert (Aloe), the temperate biome (Balsam Fir) and, according to Simlife, the tropical rain forest (Sphagnum Moss).

Go to the dashboard and select the Aloe by clicking on it. Click the Windows menu, the Biology Lab window and Edit. All of the information concerning Aloe will appear. By clicking the various buttons, the genetic character (genotype) can be changed which will result in changes in the physical expression (phenotype) of the organism. We **aren't** going to change anything except for one characteristic - mutation rate. First, click on the icon of the unzipping DNA molecule and change the mutation rate to zero, then click "make it so." Close the window. Then click Balsam fir on the dashboard, go to biology window and edit and change the mutation rate for Balsam fir to zero, click make it so. Do the same for Sphagnum moss. Save your world. Consider how powerful Simlife is in illustrating many, many important biological principles!! While still in this window, for Sphagnum moss, click the icon at the bottom that looks like an opening zipper (actually a replicating molecule of DNA). This allows even more editing choices as even more genetic information is shown. Look at the information for sprout moisture and sprout temperature. This gives you some big clues as to where in your world Sphagnum moss is going to be successful. Click "Never mind" to leave this window. Do the same for Aloe and Balsam fir and you can guess as to what the results of your experiment will be.

It's time to place the seeds. We are going to place 10 seeds of each plant in each one of our 27 boxes (30 seeds in each box, 270 total seeds for each species, a total of 810 seeds in all).

To the extreme right of the dashboard is "Help" and right above it is a red button (it should be red). This is the "Pause" button. Make sure the game is paused (the button is red). The game can be paused at any time by clicking this button. However, we are going to run this experiment automatically, stopping after 2 years, 4 years, 6 years, 8 years and 10 years. You will collect data after each pause of the experiment. Let's populate the world.

On the dashboard, click the first plant, Aloe. Note that its name appears in the "Selected Species" box. Now click the "Life tool," the icon of a DNA molecule just below the "Selected Species" box and hold the mouse button down. Scroll down this small box and highlight the "Populate" choice. You can now place 10 seeds of Aloe into the boxes of your world. Place the Aloe seeds at the top of each box, in two rows. If you miscount, go back to the “Life tool” icon, highlight "Smite" and click on the seed you want to remove. Make sure you go back to "Populate" again to continue to place your seeds. Once you have placed 10 seeds of Aloe in each of the 27 boxes, highlight Balsam fir, go to the "Life tool" icon, highlight "Populate" and place 10 seeds of Balsam fir in each box. Do the same for Sphagnum moss (note that **each seed of Sphagnum moss appears as 9 dots**, so each group of 9 dots is actually 1 seed). **SAVE YOUR WORLD!!**

We are finally ready!!! Go to the Simulation menu, Technical submenu, click "Run Control." Set the time to 2 years, click "Make It So." Click the pause button (the red light goes out) and your experiment begins!! In the upper left of the dashboard, time can be seen passing. After two years the simulation stops - a large message box appears on the screen. At this time collect data: **count the number of seeds for each species that has germinated (sprout)**. After the data have been collected, hit the pause button again and the simulation will run another 2 years (4 in all, as shown in the time window at the upper left) **automatically**. Collect data, run the simulation again, etc., etc. Again, collect data after 2,4,6,8, and 10 years. **SAVE YOUR GAME AFTER IT RUNS FOR 10 YEARS BY CLICKING "SAVE AS." SUBSTITUTE THE OTHER PARTNER'S NAME IN THE FILE AND THEN CLICK OK!!** (You will now have 2 files saved on your floppy disk).

Data after 2 years:

Species	Wet	Moderate	Dry	Wet	Moderate	Dry	Wet	Moderate	Dry	Temperature
Aloe										HOT
Balsam fir										HOT
Sphagnum moss										HOT
Aloe										MODERATE
Balsam fir										MODERATE
Sphagnum moss										MODERATE
Aloe										COLD
Balsam fir										COLD
Sphagnum moss										COLD
	DEEP		SOIL	MODE	RATE	SOIL	SHAL	LOW	SOIL	

Data after 4 years:

Species	Wet	Moderate	Dry	Wet	Moderate	Dry	Wet	Moderate	Dry	Temperature
Aloe										HOT
Balsam fir										HOT
Sphagnum moss										HOT
Aloe										MODERATE
Balsam fir										MODERATE
Sphagnum moss										MODERATE
Aloe										COLD
Balsam fir										COLD
Sphagnum moss										COLD
	DEEP		SOIL	MODE	RATE	SOIL	SHAL	LOW	SOIL	

Data after 6 years:

Species	Wet	Moderate	Dry	Wet	Moderate	Dry	Wet	Moderate	Dry	Temperature
Aloe										HOT
Balsam fir										HOT
Sphagnum moss										HOT
Aloe										MODERATE
Balsam fir										MODERATE
Sphagnum moss										MODERATE
Aloe										COLD
Balsam fir										COLD
Sphagnum moss										COLD
	DEEP		SOIL	MODE	RATE	SOIL	SHAL	LOW	SOIL	

Data after 8 years:

Species	Wet	Moderate	Dry	Wet	Moderate	Dry	Wet	Moderate	Dry	Temperature
Aloe										HOT
Balsam fir										HOT
Sphagnum moss										HOT
Aloe										MODERATE
Balsam fir										MODERATE
Sphagnum moss										MODERATE
Aloe										COLD
Balsam fir										COLD
Sphagnum moss										COLD
	DEEP		SOIL	MODE	RATE	SOIL	SHAL	LOW	SOIL	

Data after 10 years:

Species	Wet	Moderate	Dry	Wet	Moderate	Dry	Wet	Moderate	Dry	Temperature
Aloe										HOT
Balsam fir										HOT
Sphagnum moss										HOT
Aloe										MODERATE
Balsam fir										MODERATE
Sphagnum moss										MODERATE
Aloe										COLD
Balsam fir										COLD
Sphagnum moss										COLD
	DEEP		SOIL	MODE	RATE	SOIL	SHAL	LOW	SOIL	

Total plants present after 10 years:

Aloe _____
Balsam fir _____
Sphagnum moss _____

Based on your data, answer the following questions:

1) List in order (from most successful to least) the five sets of climatic conditions (e.g. deep soil, hydric, cold, etc.) in which Aloe grew the best. Do the same for Balsam fir and Sphagnum moss.

2) Based on this experiment and the knowledge of the habitats in which these plants grow, did any of them grow in habitats that you would not expect?? If so, why?

3) What might be the effect on seed germination if water availability is reduced for each of. the three plants? Consider the plants separately

4) Compare your results with two other pairs of experimenters. List their names and compare your results with theirs. Were they similar or different? If they were different see if you can determine why? Was there error in setting up the experiment? Were there slight variations in the environmental conditions that were established? Explain in detail.

5) Go back to the computer. Turn it on, load in the world you saved after the 10 year run of the experiment. To review, when Simlife comes on it will have a "Load a Game" bar at the bottom. Click this bar, then click the "a" drive box, which will access the files on your floppy disk. Click and highlight the second file saved and click "O.K.."

Now click the Windows menu and the "Census" submenu. Highlight "Population." Check your last results with the results in the population box. Do they agree? If not, why?

From the Census submenu click and highlight the "mortality choice." It will tell you why each species died, and how many of each species died for the reasons listed. Copy these data.

Again, from the Census submenu, click on the graphs choice. Roughly sketch the graphs below (label the graphs, please).

Lastly, from the Census submenu, highlight history and you will be able to read the complete history of your experiment. Make notes of any important information that you haven't already included.

Name________________________________

Exercise 16: SIMLIFE #3-The Scenarios

We've done two exercises using Simlife: the tutorial, which introduced the program and a detailed experiment where we observed the effects on seed germination from varying the physical parameters of an ecosystem. Now for a little fun.

There are built-in scenarios in Simlife, ecosystems that have already been constructed. The physical, chemical and biological parameters have all been set and all you have to do is observe what is going on, record the data periodically, and as the scenario runs, manipulate some of the physical and biological parameters of the ecosystem. There are six built-in scenarios in Simlife and while we are only going to run one of them, **Desert to Forest**, the student is encouraged to try them all at some point. You may even want to purchase the program for yourself. After running this scenario, in the time remaining, you will have a chance to design your own ecosystem.

"Desert to Forest" is a simulation designed to show ecological succession. While the student will minimally have to change the settings, observing and recording the information in the dialogue boxes will be of utmost importance, and a number of questions will have to be answered.

First turn on the computer. At the "C" prompt, type "Win" to start Windows and when the "Program Manager" screen appears, click the Maxis icon and then the Simlife icon to begin the program. When the introduction is finished and the last screen appears, click on the "Desert to Forest" button and read the information in the dialogue box. Now click the "play scenario" button.

The world is built before your eyes including the community of organisms. Note the rectangle in the "Map" screen. That is where you'll be starting. Now close the map screen (click on the little minus sign in the upper left hand corner). and maximize the "Desert to Forest" screen.

Note that the pause button is **on** (the scenario isn't running). This is the red button in the upper right of the dashboard. Don't touch it yet. Before we run the scenario we should add some pollinating insects to the world so that the vegetation can survive.. Click on the "population" button and click the "selected species" button. When the animals appear, click on the yellow jacket. Add ~100 yellow jackets, **scattered**, **on land** and click "make it so." Now push the pause button and our scenario begins.

Keep an eye on the dialogue box in the upper left, it shows what is happening. Also note the time passing. After about 3 years (more or less), hit the pause button to stop the scenario. We now gather the data available to find out what has been going on. First let's see what plant species we are dealing with.

In the upper left of the dashboard, click the "P" which is under a left and right arrow. This will show the plants that are present in this world. Click on each plant with the mouse and the dialogue box in the center will identify the specific plant. List the plants in the table below. Make sure you hit the right arrow button to show all of the plants present, they will scroll across the screen.

Species	3 Years	5 Years	10 Years	15 Years	20 Years	25 years	35 Years	50 Years

Now, go to the "Windows" menu selection and click on "Census," and then "Population." Click on each species. The name of each plant will appear in the center dialogue box and its population count will appear in the top right of the "Population" window. Record the population of each species. Now, within the "Population" window, click the icon that has a "down, black arrow" above a "brick." Then click on "History." A dialogue box appears which shows the time passed and what significant events have occurred in the world. In the space below, record any organisms that have become extinct. Go to the icon of each extinct plant and double click it. The "Plant Lab" appears with a description of each plant. You already know what type of ecosystem this scenario is, why have these plants become extinct? After recording these data, click the "minus" sign to close the screen and return to the "Desert to Forest" screen (don't close this screen).

Before we continue the simulation, click on the "A" button ("A" for animals) on the dashboard. What is the only animal present in our world? Find out how many of this species exists. If the number is zero, click the "Populate" button in the upper right of the dashboard and place 100 of these organisms, on land, scattered, in this ecosystem.

Go back to the "Desert to Forest" screen, click the "Pause" button to continue the scenario.

After 5 years pause the scenario again. Fill in the table as you have previously done--click on "Windows," "Census," and "Population," record the number of each species present. Below record the species that are extinct. Double click each of these plants to get to the plant lab and see if you can theorize why each of these plants have disappeared. Close all screens, except "Desert to Forest."

Run the scenario pausing after 10, 15, 20, 25, 35 and 50 years recording the same information as you have previously done. Save your simulation to a floppy disk.

What has been the overall trend of the populations during this scenario. What species are thriving? Why? Draw some conclusions about ecological succession from this scenario.

Now we are going to begin this scenario again, from time 0, however we are going to change some aspects of the physical environment before we run the simulation.

After beginning the "Desert to Forest" scenario, making sure the pause button is still on, close the "Map of the World" screen and maximize the "Desert to Forest" screen. We are now going to change the world a little bit.

First, click on the "moisture tool," which is the drop of water in the center of the dashboard. Click the up arrow to its left and increase the amount of moisture in the upper left of your "Desert to Forest" screen, and in the lower right of the screen. The boxes that have appear on your screen when you clicked on the moisture button turn dark green as the moisture level is increased. Now click this button off. Click on the "temperature" button and with the down arrow, decrease the temperature in the lower left and upper right of the "Desert to Forest" screen. Make sure there aren't any large water bodies where the world is being changed, if there are, just move to an area where there is mostly land. Click the "temperature" button off. Record how the environment has changed.

Hit the "Populate" button in the upper right of the dashboard. Choose "All Plants" from the selected species. Choose 40 as the number, choose "Scattered," and "On Land." Click "Make it So." Do the same thing for the one animal species in this scenario.

We are ready. Hit the "Pause" button and start the scenario. Keep track of the events in the dialogue boxes. Stop the scenario after 3 years and record all of the data in the table below. Do the same as in the first part of the exercise after 5, 10, 15, 20, 25, 35, and 50 years, as well as 75 and 100 years!!! Record all of the results!!! Save the scenario to the floppy disk.

Compare the results from the first run of this scenario to the second. In a carefully constructed essay, using the concepts of ecology that have been the foundation of this course, explain what has happened to this ecosystem over time both before and after the physical parameters of the ecosystem were altered. Why have these changes occurred? What would happen if the simulations were run for another 100 years? Using the manipulation tools of Simlife, how could each ecosystem be perpetuated indefinitely?

Species	3 Years	5 Years	10 Years	15 Years	20 Years	25 Years	35 Years	50 Years	75 Years	100 Years

In the time that is remaining, see if you can design your own ecosystem and make it successful. Start Simlife over and at the opening screen, choose "Experimental Mode." Choose the size of your world as "Tiny" and make it so. Here are a few ideas as to what to do: Design a specific biome-use the modification tools of Simlife and create specific environmental conditions to approximate a biome (e.g. tundra, boreal forest, temperate forest, tropical rain forest). Populate this biome with species that will survive in the biome of your choice. Change some of the organisms that are available, using the "Biology Lab," so they will be successful. Save your scenario to disk **before** you actually run the scenario (before you hit the pause button). Show your instructor the world that has been created. Run the scenario, see if it is successful.

Another idea is to illustrate some important concept of ecology: Create a predator-prey relationship. The clue here is to keep the world small and observable. Or, for example, test the law of tolerance. Create a world of extremes from one end of the system to another. Place species throughout the world and see how each responds to variations in the environmental factors. Make up your own experiment. In the space below indicate what you are going to do, how you're going to do it, the eventual results, what you predict and what your conclusions are after the scenario has run for 75 years (or until your scenario has failed, tsk, tsk).

Purpose:

Construction of the System:

Predictions:

Results:

Conclusions:

Name______________________________

EXERCISE 17: DANGEROUS PLANTS AND ANIMALS I

We are fortunate that many of the dangerous plants and animals that plague populations of under-developed countries are not present in the United States. For many, their vectors are simply not native to North America. Furthermore, our high standard of living, sanitary treatment of waste water and medical technology eliminates these organisms from our environment. Nonetheless, it is important to study these organisms. There have been reported cases of some of these organisms in this country, there are dangerous organisms native to the United States and some parasitic diseases are on the increase.

For the next three exercises, these organisms as well as a variety of medicinal plants, will be observed in the laboratory. The student will make note of the discussions of these organisms on the following pages, add other information that the instructor mentions concerning these creatures and **MAKE SKETCHES OF ALL THE MATERIAL OBSERVED!!!!**

Make note of the instructions concerning each organism such as the magnifying power to be used. Copy down any further instructions that are given and be sure to handle the microscopes properly.

"DP&A" I:
1) Phylum Protozoa (Unicellular organisms)
2) Phylum Cnidaria (Jellyfish, corals, etc.)
3) Phylum Platyhelminthes (Flatworms) including slides of the life cycle of *Schistosoma*
4) Phylum Aschelminthes (Roundworms)
5) Phylum Echinodermata (Sea Urchins)

"DP&A" II
1) Phylum Arthropoda-Class Arachnida
2) Phylum Arthropoda-Class Chilopoda
3) Phylum Arthropoda-Class Insecta

"DP&A" III
1) Phylum Chordata-Class Vertebrata-Order Chondrichthyes (Fishes)
2) Phylum Chordata-Class Vertebrata-Order Reptilia (Snakes)
3) Dangerous Bacteria and Plants
4) Medicinal Plants

I: Phylum Protozoa-unicellular organisms.

A. 1. *Trypanosoma gambiense* - African sleeping sickness. These organisms are under the demonstration scope that the instructor sets up.

2. The vector (carrier) is the Tse tse fly (*Glossina*), observed both as a slide and in the plastic mount. This disease is still important in parts of equatorial Africa as a threat to both humans and cattle. It is important to note that because some areas of Africa are so infested with this disease, human habitation is small and this lack of human activity has kept these areas is a **pristine (undisturbed) condition!!!**

B. 1. The *Anopheles* mosquito is observed under low power. This is the vector for the most serious parasitic disease in the world today. Malaria affects millions of people and one form of the disease is fatal. The other types are chronic diseases that can last a lifetime. Scan the slide so that the entire organism can be drawn.

C. 1. Observe *Trichomonas vaginalis* under high power. This common cause of vaginitis results in general discomfort, itching and a discharge. Males can get this infection of their urogenital tract.

D. 1. Observe the demonstration scope of *Entamoeba histolytica.* This organism causes the very serious, amoebic dysentery.

2. Observe the demonstration of *Giardia lamblia.* This parasite, which causes dysentery, is cosmopolitan and is becoming a serious problem in municipal drinking water in the United States.

II. Phylum Cnidaria - Jellyfish, Corals, etc.

A. 1. Observe coral specimens in plastic. How are they dangerous?

B. 1. Observe specimens of *Aurelia*, the jellyfish, in plastic. Why is this organism dangerous to humans?

C. 1. Look at *Physalia* in plastic as well as the preserved specimen. The common name for this organism is the Portuguese Man-Of-War. They are found in the tropics and subtropics and have stinging tentacles up to 50 feet long.

III. Phylum Platyhelminthes-the flat worms (Platy=flat, helminth=worm).

A. 1. Draw a sketch of the Human Liver Fluke, *Clonorchis sinensis*, under low power. Scan the entire slide so the entire organism can be drawn.

B. 1. Draw *Fasciola hepatica*, the sheep liver fluke, under low power. Follow the instructions for the previous slide. Observe the preserved specimens in the bottle.

C. 1. Draw the **scolex** of the tapeworm, *Taenia pisiformis* under low or medium power. This is the organ by which the tapeworm attaches to the small intestine of its host.

C. 2. Tapeworms can grow up to 60 feet in length. Some tapeworms do nothing more than cause intestinal obstructions. Others, however, can absorb important vitamins and nutrients resulting in serious illness. Tapeworms are usually contracted by drinking and eating contaminated water and food, especially pork and beef.

D. 1. Sketch all the stages in the life cycle of the blood fluke, *Schistosoma*. The instructor will set up demonstration scopes for the **egg, miracidium and cercaria**. The student will observe the adult male and female under medium power on their individual scopes.

D. 2. Where in the world are these creatures found?

D. 3. What is the world-wide significance of these parasites?

D. 4. Describe the life cycle of these organisms.

IV. Phylum Aschelminthes - the roundworms.

A. 1. Draw a sketch of the hookworm, *Necator americanus* under medium power. This organism, brought to this country from Africa by the slaves, lives in the small intestine where it sucks blood. Hundreds of these worms will cause severe problems. The worm is contracted by walking barefoot. The larval worm will burrow through the skin, into the blood vessels where it will travel until it reaches the small intestine. Under high power, look at the head region where "teeth" may be observed. This is how the worm attaches to the intestinal wall and sucks blood.

B. 1. Look at the specimens of *Ascaris* in plastic as well as the preserved specimens in the jar. How is this organism dangerous? How is this organism contracted?

C. 1. Under medium or high power, sketch *Trichinella spiralis*. This worm causes trichinosis. Why is this disease dangerous?

C. 2. Now sketch the slide of the worms embedded in muscle tissue.

D. Filarial worms are microscopic and deposited into the blood stream by different vectors. Below are three filarial worms which are of some significance. It has been found that the treatment for heart worm in dogs, is also effective treatment for river blindness, *Onchocerca volvulus*. Since the pharmaceutical company which makes the drug, Ivermectin, is donating it to the west African and Latin American countries that suffer from river blindness, it is believed that this disease will soon be under control and no longer be the terrible problem that it presently is.

D. 1. Draw a sketch of the slide of river blindness, *Onchocerca volvulus*. This worm is carried by the black fly, *Simulium damnosum*. After being bitten, a bump will form on the head of the victim. The bump is filled with worms that eventually burrow through the eye!!, causing blindness. In some primitive villages, almost all of the inhabitants are blind!!!!

D. 1a. Draw the slide of the blackfly, *Simulium damnosum* under low power. Scan the entire slide.

D. 2. The filarial worm, *Wuchereria bancrofti*, is under a demonstration scope. This slide is a blood smear so many red blood cells will be observed as well as the worms.

D. 3. The worm is carried by a mosquito and is a disease in the most backward and underdeveloped countries. After years of being bitten by mosquitos, and having thousands and thousands of worms in the circulatory system, the worms eventually block the lymph glands and cause lymph to accumulate in these areas. These areas of the body become grossly swollen and distended resulting in a condition called elephantiasis. Approximately 100 million people suffer from this disease today.

D. 4. Draw a sketch of the dog heart worm, *Dirofilaria immitis*. Although this isn't a disease of humans, a dog, after being bitten by mosquitos, will eventually have worms in its heart, killing it. So make sure your dog gets heart worm pills.

E. 1. Draw a sketch of the pinworm, Enterobius vermicularis under medium power. Although this worm is essentially harmless, what problems does it present?

V. Phylum Echinodermata (Echino = spiny, dermata = skin) - the starfish and sea urchins.

A. 1. Observe the specimens of *Arbacia* in plastic. Obviously, why is the sea urchin dangerous to humans?

Name________________________

EXERCISE 18: DANGEROUS PLANTS AND ANIMALS II

I. Phylum Arthropoda (Arthro = jointed, poda = foot).
Class Arachnida-spiders, ticks, mites, etc.

A. 1. Observe the tarantula in the plastic mount. Although all spiders produce venom, the native tarantulas are not very toxic to humans and are even reluctant to bite humans.

B. 1. Observe the preserved specimens of *Lactrodectus mactans*, the black widow spider. This creature can inflict a very serious, and sometimes even fatal bite. Its venom is very toxic. These organisms live in remote, out-of-the-way places around your home. Note the hourglass-shaped marking on the spider. This is the most common diagnostic characteristic.

C. 1. Draw a sketch of the preserved scorpions. Where is the venom apparatus? What is the habitat range of these creatures in the U.S.?

D. 1. Draw the wood tick, *Dermacentor andersonii*, under low power. This creature is the vector for Rocky Mountain Spotted Fever.

D. 2. What is the habitat of these creatures and how are they removed from the body?

E. 1. Draw a sketch of the deer tick *Ixodes dammini*, the vector for Lyme Disease. This disease is easily treatable with antibiotics, which will minimize the effects of the disease, if detected early. But since detection many times occurs when the disease is already running its course through the body, many serious conditions can result including arthritis and persistent fever and weakness. As you see, the tick is extremely small and difficult to detect on your skin. There are effective tick repellants available.

F. 1. Draw a sketch of the itch mite, *Sarcoptes scabeii*, under medium power.

F. 2. This organism causes a skin disease in humans which is not very serious, but in fur-bearing mammals, it causes the mange. This is a slow, painful, agonizing and fatal disease to these animals.

II. CLASS CHILOPODA-Centipedes

A. 1. Draw a sketch of the preserved specimen of the centipede. Although they can inflict a painful bite, New Jersey centipedes are basically harmless. Southwestern and tropical species are not only larger, but can have extremely painful bites that are sometimes fatal.

III. CLASS INSECTA-The insects are the most successful creatures on this planet. They evolved much earlier than humans and will be here millions of years after we, as a species, disappear.

A. 1. Draw a sketch of the human louse, *Pediculus humanus*, under low or medium power.

A. 2. Draw a sketch of the pubic, or crab louse, *Phthirus pubis*. This organism is contracted from contaminated bedding or from someone who already has it who you have come in contact with.

B. 1. Observe and draw the life cycle of the screw worm, *Callitroga hominivorax* in the plastic mount. This organism represents a severe economic problem to cattle ranchers. The adult female fly lays her eggs in the ears, nose and open cuts of cattle. When the maggots hatch, they burrow into the cattle. A large infestation can kill the cow or steer.

C. 1. Observe and draw the slide of the bedbug, *Cimex lectularius*, under low power. Although it can be a vector for various diseases, it is mostly a pest feeding off humans in infested bedding, mattresses, etc .

D. 1. Draw a sketch of the rat flea, *Xenopsylla cheopis*, under low power. This is the vector for the bubonic plague. What was the significance of this disease?

E. 1. The housefly, *Musca domestica*, can be the vector for many diseases. Draw sketches of the life cycle, found in the plastic mount.

F. 1. Draw the different species of cockroaches in the plastic mount. Although this organism can be the vector for many diseases, it is mostly a cosmopolitan pest. The cockroach that is most common in this region is the "German" cockroach.

G. 1. Draw the termite life history in the plastic mount. Termites have a mutualistic symbiotic relationship with protozoans.

H. 1. Draw the *Culex* life history in the plastic mount. This mosquito is the vector for many diseases.

I. 1. Draw and identify the five insects in the "Household Pest" plastic mount. Hint: You have already seen all of these organisms.

J. 1. Draw the "Insects Pests and Vectors" in the plastic mount. Label drawings with their common names. What is the significance of each organism?

NAME___________________________

EXERCISE 19: DANGEROUS PLANTS AND ANIMALS III

I. Phylum Chordata
Class Chondricthyes - cartilaginous fishes

A. 1. Draw the shark jaws and the preserved specimen of the sand shark.

A. 2. What is the most dangerous shark in the world and where is it found?

II. REPTILES

A. 1. Draw the preserved specimen of the timber rattlesnake (*Crotalus horridus horridus*). This creature is on the State of New Jersey's endangered species list.

A. 2 What is the other species of poisonous snake found in New Jersey?

A. 3. What are the two other species of poisonous snakes found in the U.S.?

III. BACTERIA

A. 1. Draw a sketch of *Staphyloccus aureus* under the demonstration scope. Staph infections can occur in many different areas of the body causing everything from acne to very serious systemic infections.

A. 2. Draw a sketch of *Neiserria gonorrhea* under the demonstration scope. Guess what disease this bacterium causes?

A. 3. The last demonstration scope has *Bacillus anthracis* under it. Anthrax is a very serious disease, originally called "wool-sorters" disease because it was contracted from sheep. Antibiotics minimize the severity of this disease if it is "caught" early enough.

IV. Medicinal Plants - Draw sketches of the 12 herbarium sheets of medicinal plants. The roots of medicine and the prescription of drugs had its basis (and still does) in the use of plants. Please take note that the medicinal properties of these plants are suspect, most based on legend, folklore, and anecdotal information although some medicinal properties have been substantiated. Make note of the brief descriptions for each plant.

1. Wild sarsparilla (*Aralia nudicaulis*) was used as a "blood-purifier," a general tonic to make a person feel better. In other words, "it cures what ails ya."

2. Witch hazel (*Hamamelis virginiana*) is an astringent used to treat hemorrhoids, treat minor pain, keep muscles from cramping, and in eye washes, etc.

3. Spicebush (*Lindera benzoin*) was used by native Americans to treat coughs, cramps and as a "blood purifier." Its berries were made into a tea.

4. Belladonna (*Atropa belladonna*) is a source of atropine which is used to treat spasms and dilate the pupils of the eyes

5. Sweet fern (*Comptonia peregrina*) is not a fern at all and used to treat poison ivy, prevent vomiting, treat diarrhea and as a general "healthful" beverage.

6. Foxglove (*Digitalis purpurea*) contains the heart-tonic glycoside, digitalis. It is used today to treat congestive heart failure - make the heart pump harder. It is an extremely important drug. In the wrong dosage, as all medicines, it can be very dangerous

7. Wintergreen (*Gaultheria procumbens*) is the source of true wintergreen oil which is used as an astringent-EXTERNAL use only. It is very toxic if taken internally!!

8. Boneset (*Eupatorium perfoliatum*) was used to break fevers especially during flu epidemics. Recent research indicates that this plant may indeed, stimulate some non-specific portion of the immune system.

9. Ipecac (*Euphorbia ipecacuanhae*) is an extremely strong laxative which is still used today. It is the drug of choice to treat some types of poisoning but must be used very carefully.

10. Peppermint (*Mentha piperita*) is used in flavorings and to treat indigestion. It may even be effective in treating Herpes simplex viral diseases. The oil cannot be taken internally because it can cause severe allergic reactions.

11. Sassafras (*Sassafras albidum*) was used as a tea and was a "cure-all." It was widely used and extremely popular. Recently, the oil of sassafras, has been reported to be carcinogenic.

12. Dogbane (*Apocynum cannabinum*) was used to treat heart ailments, warts and to induce vomiting. It is extremely dangerous if taken in the wrong dosage as it contains cardioactive substances that can adversely affect the heart.

V. Poisonous Plants. Draw sketches of the 12 herbarium sheets of these poisonous plants and make note of the discussions concerning them.

1. Poison hemlock (*Conium maculatum*) was taken by Socrates. Why?? It's Greek to me but the leaves and juices are extremely poisonous. A member of the "carrot" family, it resembles many other plants.

2. Water hemlock (*Cicuta maculata*) is very similar to the above and resembles wild parsley!!! Be very careful that you don't misidentify it.

3. Jimsonweed (*Datura stramonium*) is extremely toxic - all parts of the plant. It can cause hallucinations and toxic reactions. It does have medicinally properties but must be carefully administered.

4. Pokeweed (*Phytolacca americanum*) is a very common, tall perennial plant. While the young leaves are reported to be edible, all parts of the plant are considered toxic and should be avoided.

5. Mayapple (*Podophyllum peltatum*) was used by the native Americans as a "body cleanser'" a strong purgative. It's leaves, however, are very toxic. The fruit, which appears in the spring, and looks like a small apple (hence the name, Mayapple) is supposed to be edible.

6. Nightshade (*Solanum dulcamara*) has been used to treat some forms of cancer but is highly toxic, causing vomiting, convulsions and paralysis. It is a member of the potato family and is related to the tomato.

7. Lily-of-the-Valley (*Convallaria majalis*) is potentially toxic and can be a skin irritant. It does have medicinally uses but must be administered very carefully.

8. Tansy ragwort (*Senecio jacobaea*) was used by the native Americans as a general medicine but the entire plant can be toxic.

9. Bloodroot (*Sanguinaria canadensis*) has been shown to have a number of medicinal purposes. Most recently it has been used in toothpaste and mouth-washes to prevent plaque build up. However, an improper dosage can have toxic effects.

10. Milkweed (*Asclepias syriaca*) is a very common plant which contains many cardioactive compounds that are potentially dangerous. It has been used for a variety of therapeutic reasons.

11. Red buckeye (*Aesculus pavia*) is uncommon in the Northeast and has poisonous nuts, leaves, flowers and bark. Fatalities have been reported.

12. Larkspur (*Delphinium* sp.) does not grow in the Eastern United States but this pretty wild flower has poisonous roots and seeds.

VI. Some Other Dangerous Plants

1. Draw the plastic mount of *Toxicodendron radicans* (or *Rhus radicans*) - poison ivy. It is identifiable by its 3 shiny leaves. It can grow as a vine, shrubby plant or low-growing plant. It grows in a variety of habitats. It produces a fluid which causes blistering of the skin. Increased exposure makes one more sensitive to it. Learn to identify it and avoid it.

2. Look at and draw both species of ragweed (*Ambrosia*) in the two plastic mounts. Many people are sensitive to the pollen of ragweed which causes hayfever. Goldenrod, a plant that flowers at the same time as ragweed, does not cause hayfever. Treatment of hayfever is with antihistamines.

3. Draw the example of *Phoradendron* - mistletoe, in the plastic mount. Mistletoe is a parasitic plant on trees. It is poisonous to ingest and will cause severe intestinal upset. There is just one thing to do with mistletoe.

4. *Prunus*, the cherry tree, has poisonous roots, stems and leaves (fruit pits too). Intestinal upset results from ingesting these parts. Draw the flower in the plastic mount.

Name________________________

Exercise 20: Geographic Information Systems (GIS)

Scenario 1: As a population ecologist, you are studying wolf reintroduction into part of its former range. Three wolf packs have been reintroduced and it is your project to determine if the area represents their carrying capacity or if there are more resources for an increase in the wolf populations. You must pinpoint their locations and track their movements to better understand how they can be managed. Is there enough habitat? Will overcrowding cause dispersal of wolves? Can they disperse to other areas? What are the present territories for the three packs?

Scenario 2: The state is attempting to site a sanitary landfill within your town. As an environmental commissioner, you must make recommendations as to the safety and applicability of such a project. There is a great deal of protest from the residents. How can you determine if the siting of such a landfill is safe? Will the soils support a landfill? What is the impact on streams, wetlands, wells, etc. Can you gather enough data to make a responsible and accurate recommendation?

Scenario 3: As the director of planning for your town, you must make sure that increased development, i.e. new housing, commercial buildings such as malls, industry, new infrastructure, such as roads, bridges, etc., are accomplished with maintaining the semi-rural nature of your community. You are faced with the prospect of increased development, which is necessary to create a viable tax base, which results in less of a tax burden for the residents, and yet maintain the quality of life for which your town is known. Development, up to this point, has been scattered throughout the township with no "center" or hub for this development. A "Town Center" is planned which will restrict heavy development, with increased densities of housing, commercial buildings and light industry to a portion of the town and protect the environs which surround this "center." Furthermore, a "greenbelt" is proposed that encompasses the "Town Center" and includes bicycle paths, walkways, etc. How do you plan such an endeavor?

Scenario 4: As a wetlands delineator, you have been hired to ascertain the presence and absence of wetlands and to delineate them from the surrounding upland areas on a 150 acre parcel of land. How do you get a general idea of where the wetlands are located on the property?

Scenario 5: As the municipal engineer, you must work with state and federal officials to design a sewer system for your town. This will eliminate the further use of individual septic systems. How do you decide where the sewage treatment plant will be located and how and where the sanitary sewer lines will be constructed?

Scenario 6: As an ecologist, you are conducting a long term study on nutrient flow in and out of a forested watershed. This is important because these nutrients may flow into reservoirs that the watershed empties into. How do you monitor the flow of nutrients from all of the streams that flow into the reservoirs? How do you correlate soil types, with water filtration and flow? How will you keep track of all of the physical characteristics that comprise this watershed and monitor the continual inputs and outflows of nutrients?

Scenario 7: As a forester, you must analyze the land holdings of your company and design a master plan for the next ten years to insure that there will be a continual harvest of mature trees. You must decide what areas can be harvested and when and what areas should be left untouched.

Scenario 8: You are the director of marketing for the Eddie Schmenke Corporation, a mail order, outdoor clothing and recreation company. After years of selling products solely through the mail (including the famous Schmenke outdoor hunting boot), the firm is finally going to open its first retail store. The New York-New Jersey metropolitan area has been targeted for this first venture. By analyzing the demographics, traffic patterns, availability of access, per capita income of the immediate area and its environs, you must designate four potential sites for the store. Later, one site will be selected from the original four. How will you accomplish this task?

The answer to all of these problems lies in referring to detailed **maps** and this is essentially what a GIS is. A **Geographic Information System** (GIS) is a computerized mapping program that can be run, on among many vehicles, a personal computer. The data contained within the system can be illustrated graphically, as in a map, but also as tables, charts, text, etc. Conversely, data included in tables, charts, text can be expressed as maps. The data can be manipulated in many, many ways, revealing the associations and relationships between, what may seem, independent and non-relational information. A series of maps can be overlaid on each other, further revealing the relationships that exist.

These maps are representations of the real world (computer simulations or models) and go beyond the limitations of flat, two-dimensional depiction.

GIS allows information to be presented simply and clearly and allows these data to be visualized in new ways to give new perspectives.

GIS, like any other computer program is only as good as the data entered into it. But there is already a wealth of data available from many independent sources: the federal government, states and municipalities and private industry. These data can be purchased for use in a GIS. For example, the state of New Jersey has released three CD-ROMs of data concerning the state. Maps and data of municipalities, geology, wetlands, streams, lakes, flood plains, toxic waste sites and even hemlock and cedar forests are included on three discs, divided into the northern, central and southern portions of the state.

Data can be generated for use in a GIS. Maps can be created from tabular data or for example, using the United States Geological Survey Maps (USGS Topographic Quadrangles) and a hand-held GPS (Global Positioning System). A GPS is a small electronic receiving device, about the size of a large cellular phone, which receives signals sent from satellites orbiting the earth. By receiving a number of signals, the exact latitude and longitude of the point where the GPS is receiving those signals is indicated on the display. It can have an accuracy to a few feet. These

GPS instruments are the peace-time applications of missile and rocket technology which allow targets to be located and destroyed.

An example of a GPS used with a GIS would be a study to locate all the known denning sites of the timber rattlesnake (*Crotalus horridus horridus*), a state protected species, in northern New Jersey. An investigator would visit known sites or find new sites and take readings off the GPS. These would then be plotted on a USGS map or a GIS map, such as the ones available from the New Jersey Department of Environmental Conservation.

The GIS exercise here, uses[1]**ArcView®™**. ArcView is a state-of-the art GIS software program, widely used, and is a standard in the field. Many data sources are originally constructed in ArcView format because of its wide use and flexibility. ArcView allows data to be easily accessed, displayed, manipulated and published. It can link other data formats and incorporate them into ArcView's.

For today's exercise, GIS and ArcView are going to be introduced in a multimedia presentation. It is impossible to even teach the basics in a three hour lab. The object of this exercise is to illustrate how powerful GIS is and that it is state-of-the-art technology with many applications as illustrated in the scenarios previously presented. The object is to expose the student to these applications and spur an interest that might later be pursued in greater depth.

This computer presentation is entitled, "Getting to Know Desktop GIS." It runs itself except when subsequent sections must be accessed. This is accomplished by a click of the mouse.

Start Windows and click the "Getting to Know GIS" icon. First view the part called, "Desktop GIS Primer." There are six sections to this. Access the three sections on the left and then the three on the right. The second part is called, "ArcView Showcase." Again, access the three sections on the left and then the three on the right.

After viewing this presentation and drawing from the semester's discussions of ecological and environmental problems, create your own scenario for which GIS would be important in addressing the issues.

---

[1] ArcView is a registered trademark of Environmental Systems Research Institute, Inc. (ESRI). 380 New York Street. Redlands, CA 92373-8100.

Name____________________________

EXERCISE 21: A FIELD TRIP

This is a note to both student and instructor. This lab manual was designed to allow students to study the environment around them vicariously, without leaving the class room. However, there is no substitute for getting outside and doing some field work. In an urban setting, this is very difficult, if not impossible. In a suburban setting it is equally difficult but in other ways. Having 25 students tromping around the field is very unwieldy. The various levels of outdoor experience as well as the various levels of enthusiasm from primarily non-science majors can result in disaster. Furthermore, the environment of suburbia may not lend itself to field work. At Bergen Community College there is very little in the way of "natural habitat" on campus. Although some county parks are close by, the logistics of getting the class to these sites is difficult. Keeping the class together, is another story.

It is suggested that a field trip be attempted. If it is impossible, well, this last lab can be ignored. The following lab is based upon a very short outdoor experience that we have been doing on this campus for quite a long time. It always varies in its success. For the instructor, feel free to modify it to make it relevant to your particular situation.

The purpose of this field trip is to simply expose the student to the flora and fauna of a typical suburban environment.

We start our field trip behind the tennis courts where a number of gray birch (*Betula populifolia*) are growing. The interesting point is that these trees, which are successional or transitional species (not "climax"), are reproducing both sexually and asexually. Note the fruit, which is called an ament or catkin. Also note that the tree is sending up above ground shoots from the base of the trunk, this is asexual reproduction.

Note the "weeds" growing in this exposed spot. We find ragweed (*Ambrosia artemisiifolia*), poison ivy (*Toxicodendron radicans*), mugwort (*Artemisia vulgaris*) and a variety of other species.

What are some of the other species found at this site? What is the significance of ragweed?

Note the specimens of scarlet oak (*Quercus coccinea*). This is a tree found in a mature forest setting. Note the transitional species found, sassafras (*Sassafras albidum*), black cherry (*Prunus serotina*), etc. Why are these species so large? What shrubs and herbaceous species are found at this site.

Black locust (*Robinea pseudoacacia*) is a species that is happier south and west of New Jersey but is doing very well at this site. Note the thorns on the branches.

At the "E" pond please note the eutrophic condition of this ecosystem. This pond is teeming with life and many of the classic microscopic organisms that are studied in general biology can be found at various times of the year here.

Observe the plants growing around the margins of the pond. The most significant species is purple loosestrife (*Lythrum salicaria*). This plant is an exotic species, not native to North America. It is doing very well here. As a matter of fact, it is doing too well here. It is displacing the native species. Despite its beautiful purple flowers which bloom in late July and early August, it is not an important resource to wildlife. Cattail (*Typha latifolia*), which it is displacing, is an important resource. As you pull into the campus, note the small wetland on your right hand side, adjacent to the golf course. Twenty years ago it was a small cattail marsh. Now it is approximately 99% purple loosestrife.

What other species are encountered around the pond and its environs?

After doing the wetlands delineation lab exercise, we can actually see how a delineation is done in the field. What is the Munsell Color of the soil? What are the plants in the wetland adjacent to the pond? Is there any evidence of wetland hydrology?

List the plants encountered on the walk from the pond to the "E" parking lot.

This small grove represents the "wildest" portion of our campus and contains species found in a mature forest ecosystem. Note the following species: red oak (*Quercus rubra*), white oak (*Quercus alba*), American beech (*Fagus grandifolia*), American elm (*Ulmus americana*), the shrub, witch hazel (*Hamamelis virginiana*), three species of fern, New York (*Thelypteris noveboracensis*), cinnamon (*Osmunda cinnamomea*) and marginal shield fern (*Dryopteris marginalis*). Note stinging nettle (*Urtica dioica*) which is also common around the pond. Note Virginia creeper (*Parthenocissus quinquefolia*), mayapple (*Podophyllum peltatum*), and the other species. List them.

On our walk back, note another species of the mature forest, hickory (*Carya* sp.). Also note fox grape (*Vitis Labrusca*) and list the other species encountered.

In class, classify all the species as trees, shrubs and herbs. Make notations of the trees as to whether they are transitional (successional) species or if they are "climax" species. Annotate all the other species with facts derived from this field trip. Use the back of this sheet to complete this exercise.

APPENDIX A: LIST OF MATERIALS FOR EACH EXERCISE AND ANNOTATIONS FOR THE INSTRUCTOR

EXERCISE 1: The Microscope

Compound microscopes
Lens paper
Slides of letter "e," *Amoeba, Paramecium, Vorticella*
Blank slides, cover slips
"Hay infusion"
Toothpicks
Methylene blue stain
Elodea

This is a fool proof method and will allow even the most inept to use the microscope. Stress the need for students to ask the instructor questions rather than their lab partners. This lab doesn't run for three hours so additional material may be used and/or other exercises.

EXERCISE 2: Introduction to Ecology

Aquarium
Video: Race to Save the Planet, Part 1 or some other relevant video

A basic review of ecological principles plus a discussion of methods including "observation," would be desirable. I usually give them more of the "compare the top picture to the bottom." There are many good videos out now stressing basic ecological principles.

EXERCISE 3: Pond Study

Prepared slides of *Anabaena, Oscillatoria, Spirogyra, Oedogonium*, mixed Diatoms, *Amoeba, Paramecium, Stentor, Vorticella*, Hydra, Planaria, Rotifers, Cyclops, Crayfish.

Clean slides, cover slips
Pond water containing vegetation or debris.

Any pond will do, the more eutrophic, the better. Sampling at different times during the year results in a greater diversity of organisms, except in the winter. In an urban setting where access to a pond may be difficult, a fish tank may provide some live material.

EXERCISE 4: Paleoecology

Refer to the citations in the lab. A great deal of material can be downloaded from the 'net, including all of the material in the lab. I use a pollen diagram browser called Siteseer, that was downloaded. The paleoclimatology home page has so much material. It can be explored for weeks. This is a good lab because they do not know that this discipline even exists!!

For the construction of the pollen diagrams: Siteseer allows for the downloading of the raw pollen data in ASCII. Import these data into a spreadsheet or word processor and then reformat so that each column is well delineated.

EXERCISE 5: Ecological Modifications in Plants and Animals

Cross-section slides of Lilac (*Syringa*), *Potamogeton, Yucca*, leaf slide of hydrophyte, mesophyte, xerophyte (all three on same slide), *Typha, Pinus, Zea, Ammophila, Myriophyllum.*

I really like this lab for many reasons. They are learning some classical biology (leaf morphology) and can do it in a relatively short period of time. They then have to apply this knowledge in determining the habitats for the unknowns. **What I have found for this lab is essentially pertinent for many of the other labs: the right answers to the questions are not as important as getting answers that are well thought out and logically based upon the material observed during the lab.** Go over the lab at the end of the period and give them the correct answers.

EXERCISE 6: Ecological Sampling-a Forest Ecosystem

"Forest" boards
Meter sticks
Rulers
Calculators

For this lab and the next, the forest and meadow have been artificially created on oak-tag paper. Contact me at Bergen Community College for instructions as to constructing your own. This lab is great. Not only do the students have fun collecting the data but the statistics, although initially daunting, are within the student's capabilities. They don't like doing the statistics but they learn that ecology is more than just running around outside having fun. I'm very proud of this lab and the following one.

If possible, show 2x2 slides of the trees discussed in the lab.

EXERCISE 7: Ecological Sampling-a Meadow Ecosystem

"Meadow" boards
Meter sticks
6x6' transparent overlays

Show 2x2 slides of the vegetation discussed in the lab. This lab works well but runs short.

EXERCISE 8: Predator-Prey Relationships

Carolina Biologicals Kit #65-1990

This lab is also great because if the instructions are followed (**and they get complicated**), two fluctuating sigmoid growth curves will be obtained after the students graph out the data. It is advisable to actually show the students where to enter the data and do the first four test runs very deliberately. Still, there always will be students who mess it up.

Wa-tor is a shareware program but its licensing for educational purposes must be purchased. It can be ordered from:

Dr. Warren Kovach
Kovach Computing Services
85 Nant-Y-Felin
Pentraeth, Anglesey LL75 8UY
Wales, U.K.

EXERCISE 9: Symbiosis

Prepared slides of crustose lichen, *Rhizobium*, cross-section of root nodule, *Trichonympha, Escherichia coli, Leishmania donovani* blood smear, *Paragonimus westermanii*, *Schistosoma mansoni* adults (males and female), *Trichura trichuris, Glossina, Simulium damnosum, Pediculus humans, Phthirus pubis.*

Preserved specimens of foliose lichen, fruticose lichen, legume root mass, termite life cycle in plastic, Spanish moss.

A word about "demonstration microscopes:" I don't teach the students to use the oil immersion lens. Everything resembles salad dressing if I do. For the material that requires this high magnification, I set the material up as a "demo" scope and have the students observe the material at the instructor's lab table.

EXERCISE 10: Shelford's Law of Tolerance I

Microbiological supplies: Petri dishes with nutrient agar, nutrient broth, loops, bunsen burners, disposal bags for hazardous wastes, test tube racks, sterile swabs, etc.

Culture broths of *Staphylococcus aureus, Bacillus megaterium, Escherichia coli, Serratia marcescens*

Index cards, U.V. light boxes, table top disinfectant, rubber gloves and goggles to taste.

See below.

EXERCISE 11: Shelford's Law of Tolerance II

See bacterial cultures above.

Nutrient broth tubes at pH of 5.0, 7.0, 9.0
Petri dishes with nutrient agar
"Sterile disks"
"Toxic" substances of bleach, formaldehyde, phenol, Listerine, table top disinfectant
Forceps in alcohol

pH setup: litmus paper (red and blue), solutions of coffee, milk, lemon juice, cola, baking soda, distilled water.

Beakers
hand-held pH meters
0.05M HCL
0.05M NaOH
pH 7.0 buffer solution

I do both of these exercises in one lab period. It shouldn't take more than three hours to do and all of the data can be reviewed the following week. It is also one less week that the students have to work with the microorganisms (which sometimes scares the daylights out of me). These labs have been modified from classical microbiology labs and a good reference is the laboratory manual by Benson (W.B. Brown, publishers, any edition).

EXERCISE 12: Wetlands Delineation

Soil Conservation Survey of Passaic County (6 copies)
Federal Emergency Management Agency (FEMA) flood plain map
USGS topographic ma, Newfoundland, N.J. quadrangle (2 copies)
National List of Plant Species that occur in Wetlands, Region I - Northeast (6 copies), published by: Resource Management Group, Inc.
P.O. Box 487
Grand Haven MI 49417-0487
Rulers

Show 2x2 slides of wetlands. The wetlands on the USGS maps are palustrine scrub-shrub swamp, Atlantic white cedar swamp, poor fen and sphagnum bog. It would be ideal if slides could be shown of these wetlands. This lab is easily modified for anywhere by simply replacing the soils map, FEMA map, topographic map and the other reference material with those relevant to your area. Where most states use the ACOE manual, it would be beneficial to have a copy of that manual as a reference.

EXERCISE 13: The Wolf - An Endangered Species

Again, refer to the references in the lab. The maps and Internet data are indicated. I show the Audubon video on wolves (the one with Robert Redford as the narrator). It is excellent.

Exercise 14: Introduction to the Computer and Simlife©

IBM compatible computers loaded with Windows. The faster the machines the better. Although the lab was written for Windows 3.1 in mind, it can easily be modified for Windows 95.

For teaching students to load from floppy disks, a timely program is STP-Save the Planet which deals with Global Warming, ozone depletion, etc. and is obtained from: Save the Planet Software

P.O. Box 45
Pitkin, CO 81241
(303) 641-5035

They will send information on site licensing, etc.

Simlife-refer to the foot note at the bottom of the first page of the exercise. Site licenses can be obtained but it may be cheaper to by copies of the program locally IF YOU CAN FIND THEM!

EXERCISES 15 and 16: Simlife #2 and #3

See above.

EXERCISES 17, 18, and 19: Dangerous Plants and Animals I, II, and III

Prepared slides (whole mounts unless otherwise indicated): *Trypanosoma gambiense, Glossina, Anopheles* mosquito, *Trichomonas vaginalis, Entamoeba histolytica, Giardia lamblia, Clonorchis sinensis, Fasciola hepatica, Taenia saginata, Schistosoma* adult male, female, miracidium, redia, cercaria, *Necator americanus, Trichinella spiralis, T. spiralis* encysted in muscle tissue, *Onchocerca volvulus, Wuchereria bancrofti, Dirofilaria immitis, Enterobius vermicularis, Dermacentor andersonii, Ixodes, dammini, Simulium damnosum, Sarcoptes scabeii, Centipedes, Pediculus humanus, Phthirus pubis, Cimex lectularius, Xenopsylla cheopis, Bacillus anthracis, Staphylococcus aureus, Neisseria gonnorhea.*

Preserved or plastic mounts: *Glossina*, coral specimens, *Aurelia, Physalia,* tapeworms, *Fasciola, Ascaris, Arbacia*, tarantula, *Lactrodectus mactans*, scorpion, centipedes, *Callitroga hominovorax* life cycle, *Musca domestica* life cycle, cockroaches, termite life history, *Culex* life history, "Household pest block," "Insect pests and vectors block," rattlesnake, shark jaws, *Toxicodendron radicans, Ambrosia aremisiifolia, Ambrosia trifida, Phoradendron.*

Compound microscopes

Carolina herbarium sheet set # 23-8850, Medicinal Plants

Aralia nudicaulis - Wild Sarsparilla
Hamamelis virginiana - Witch Hazel
Lindera benzoin - Spicebush
Atropa belladonna - Belladonna
Comptonia peregrina - sweet fern
Digitalis purpurea - Foxglove
Gaultheria procumbens - Wintergreen
Eupatorium perfoliatum - Boneset
Euphorbia ipecacuanhae - Ipecac
Mentha piperita - Peppermint
Sassafras albidum - Sassafras
Apocynum cannabinum - Dogbane

Carolina herbarium sheet set # 23-8860, Poisonous Plants

Conium maculatum - Poison hemlock
Cicuta maculata - Water hemlock
Datura stramonium - Jimsonweed
Phytolacca americana - Pokeweed
Podophyllum peltatum - Mayapple
Solanum americanum - Nightshade
Convallaria majalis - Lily of the Valley
Senecio jacobaea - Tansy ragwort
Sanguinaria canadensis - Bloodroot
Asclepias syriaca - Milkweed
Aesculus pavia - Red Buckeye
Delphinium sp. - Larkspur

EXERCISE 20: Geographic Information Systems (GIS)

The text for "Getting to Know GIS" can be downloaded from ESRI, Inc. on the Internet (http://www.esri.com) but contact them at the address in the lab for obtaining the CD-ROM. There may be a minimal charge.

EXERCISE 21: A Field Trip

The field

Save yourself the aggravation. Ten percent will love it, ten percent will absolutely hate it and the eighty percent left won't care one way or the other.